Springer Climate

Series editor

John Dodson, Menai, Australia

Springer Climate is an interdisciplinary book series dedicated on all climate research. This includes climatology, climate change impacts, climate change management, climate change policy, regional climate, climate monitoring and modeling, palaeoclimatology etc. The series hosts high quality research monographs and edited volumes on Climate, and is crucial reading material for Researchers and students in the field, but also policy makers, and industries dealing with climatic issues. Springer Climate books are all peer-reviewed by specialists (see Editorial Advisory board). If you wish to submit a book project to this series, please contact your Publisher (elodie.tronche@springer.com).

More information about this series at http://www.springer.com/series/11741

Guido Visconti

Problems, Philosophy and Politics of Climate Science

Springer

Guido Visconti
Dipartimento di Scienze Fisiche e Chimiche
Università dell'Aquila
L'Aquila
Italy

ISSN 2352-0698 ISSN 2352-0701 (electronic)
Springer Climate
ISBN 978-3-319-88078-5 ISBN 978-3-319-65669-4 (eBook)
https://doi.org/10.1007/978-3-319-65669-4

Softcover reprint of the hardcover 1st edition 2017

Printed on acid-free paper

This Springer imprint is published by Springer Nature
The registered company is Springer International Publishing AG
The registered company address is: Gewerbestrasse 11, 6330 Cham, Switzerland

This book is dedicated to Richard Goody
He ran measurements on a de Havilland Mosquito
Was a friend of Werner Heisenberg
For 50 years, I learned from his papers and books

Preface[1]

Freeman Dyson is a world-renowned physicist known also for being a skeptic about global warming. Reviewing a book on the political implications of the global warming, he reports the contempt of the majority for the opinions of another skeptic, Richard Lindzen. In the preface of a report by the Global Warming Policy Foundation, the same Dyson observes that "the public perception of carbon dioxide has been dominated by the computer climate-model experts who designed the plan." These considerations contain two important points, the first is the implicit assumption of the existence of a "climate science" and the second is the fact that, if it exists, the climate science is identified with the global warming. The existence of a climate science could be proved only if we find that it follows the "scientific method" that can be summarized as follows. We start by observing some aspect of the climate system and then we make a hypothesis to explain the observations that is used to make some prediction. These predictions are tested observationally and when we arrive to a consistent picture our hypothesis become a scientific theory. A simple example could be made having meteorology in mind. Here, the observations were the basis to develop a theory for the general circulation of the atmosphere using new concepts like potential vorticity. With the advent of the computer, the relevant equations could be numerically integrated and the forecast can now be verified each day. The frontiers of such theory today are the long-term predictions and their precision.

For the climate system, the situation is such that detailed observations are lacking (for example, for the ice age cycles) and the predictions of future climate based on the General Circulation Models cannot be verified because the data are nonexistent. A possibility (which has not been very much exploited) is to make some hindcast (that is a simulation of the past) on the reconstructed (assimilated) data of the past 40 or 50 years. Even in this case, we cannot be sure that the system in the future will behave as in the past (for example, the climate sensitivity may

[1]Dennis Bray and Hans von Storch, Climate Science: An Empirical Example of Postnormal Science BAMS march 1999 © American Meteorological Society. Used with permission.

change). We are then left with a basic difficulty that is the partial lack of observations on which to base the theory and the absence of data to verify the prediction of the same theory. In any case the climate system is not studied for its general properties rather all the efforts today are directed to the prediction for the next century. As noted by Edward Lorenz in his 1975 paper on climate predictability, a real test for the climate theory would be the simulation of the ice age cycle that was beyond the computer performance at that time as it is today. This is mostly how physics works: the recent discovery of gravitational waves is the most vivid example of the confirmation of Einstein's general theory of relativity. Similar examples can be drawn from many other fields of physics like condensed matter, etc., climate science has never advanced in the same way but rather is always in the interpretative mode except now for the prediction of global warming. The prediction can only be confirmed by the future data so that there must be other ways to confirm the theory, for example, by developing new measurement strategies for the behavior of the Earth's energy balance.

Despite the lack of a consistent scientific theory, there exist a correspondent philosophy of climate science. Some paper appeared on the subject as early as 1984 (Naomi Oreskes) and then during the late 90 those of David Randall and Bruce Wielicki or Dennis Bray and Hans Von Storch. Philosophical discussion again is very much limited to climate models and most of the earlier papers were dealing with model validation. All those first class climate scientists embarked in discussions using freely Popper or Kuhn and whatever. Bray and von Storch, following professional philosophers, even classified climate sciences among "postnormal science". According to them this should be a further state of evolution from the "normal" science introduced by Thomas Kuhn and "*addresses the issue at hand when there is a considerable amount of knowledge generated by normal science in different disciplines and there is a high degree of uncertainty and the potential for disagreement due to empirical problems and political pressure. This characterization is consistent with the present state of climate sciences.*" The early studies were discussing mostly if the models were still falsifiable when they were using parameterization and tuning. Again the confusion was total considering that, the falsification concept, was introduced by Popper in connection with scientific theories and models are not theories. Philosophy of science has something to say about climate science. After all one of the purposes of the philosophy is decide if the climate sciences are within the physical sciences or if they are like biology. In the latter case the approach could be quite different but standard about quality of data, theories and models should be established. As we say somewhere in the book, "researchers work (or should work) in such a way that the nature should constraints their conclusions. Beside scientists aim to find out the real state of the nature without never reaching it. Philosophy on the other hand deals with absolute concepts." Judging from the available material these issues are quite marginal to the debate. We have not resisted the temptation to write about these topics. We start with a very simple introduction to some of the problems climate science is supposed to study and many of them are still without a reasonable answer. Then we proceed to summarize the methodology used in the study of climate including satellite observations before going to one of the central points: the environmental modeling. This is our first stint to elementary philosophy where the concept of "pragmatic realism" is introduced and also constitutes a first example how much suspect is the terrain we enter. We discovered much later that pragmatic realism is one of the main theories invented by Hilary Putnam, the Harvard philosophical guru. However Keith Beven the inventor of the "our" pragmatic realism never mention mainstream philosophy although the two concepts can be reconciled. According to Beven "practitioners" are developing and using models that are as "realistic as possible" given the external constraints (as computing capabilities). The most liberal interpretation of Putnam's pragmatic realism is on the same direction. In other words, the theory of truth is not the real goal but rather can we model the world in such a way to make sense of it and withstand its impact? Within this framework the General Circulation Models (GCM) enters full fledged in the pragmatic realism being the only tools available to predict future climate. However, important as they are, being simply an engineering method to predict climate, they must be based on climate science . Also, to be accepted as an engineering method, models need to be tested against data and climate data are quite scarce and of dubious quality. Nevertheless there are even suggestion to regard GCM as possible means to perform crucial experimentss in the science of climate and to introduce computational techniques as its third leg besides theory and experiment. This could be equivalent to substitute the Large Hadron Collider (LHC) or the LIGO (Laser Interferometer Gravitational Observatory) detectors with a supercomputer center. As a matter of fact, immense computer resources are employed to digest the data of the above experiments but nobody think that the software in itself could be called physics.

The same chapter deals with the uncertainty issue in model predictions. Uncertainty is the main excuse the establishment uses to postpone decisions while scientists do not regard it so important because the observed changes in climate in the past decades leave very few doubts about the reality. However, uncertainty must be important for scientists because considering all the intervening conditions they should be able to indicate which are the uncertainties accompanying their prediction. Citing Popper an encouragement should be given to the simplification of theories and/or models that become in this way more testable. A problem arise with the application of the falsification criteria of Popperian tradition because models are not a theory, but rather are based on it and as someone already said they arrive at the crucial test already largely falsified.

At this point, we need to ask the question on the existence of climate science and this necessarily implies to report largely Richard Lewontin ideas in particular when the traditional vision of science progress attributed to Popper, Kuhn and so on is compared with the Marxist point of view that "scientific growth does not proceeds in a vacuum." In our case, we have plenty of indicators of the political influence like the existence of the IPCC on one side (expressions of the world government) and the oil companies on the other. The proposed creation of a few supercomputer centers could implies an obvious dependence from the governments, who own them.

Finally, a short discussion is made about the deductive nature of the models meaning that from a model we can expect some kind of data to compare with reality. This contradicts the popularity the inductive method has in science and the support it got from one of the major geophysicist of all time, Harold Jeffreys. The warning from both lines of thought is not to drift too far away from data.

There are other pretenders to the field of climate science and these are the statisticians. There are many ideas from this field that could be useful with the Bayes statistics as the main force. Bayes statistics has a very peculiar feature that makes it particularly suitable to study climate and that is the possible prediction improvement as new data arrive. That approach has been shown could be decisive for a correct prediction of future climate. There was a NASA project that employed this concept but was postponed and stripped of its main qualities and now it is just too late. We all know that global warming is for real and that there are no chances to maintain the warming below the 2° and apparently it does not make sense to maintain a park of more than 40 GCM to predict a future which is already here. A possible proposal would be to use simpler models and return to consider climate science as composed of many other problems and to continue to study the functioning of the climate system with large, complex and long-term experiments as it is done in other fields of science. Climate scientists should not lose this occasion to enter the field of "big science".

This book was written in almost solitary confinement that coincided with my retirement. A great and essential support came from Richard Goody, who read chapter after chapter and provided comments that have been included here. At the end of each chapter, a box reports a discussion between a "humanist" and the "climate scientist" on the content of that chapter. The idea, probably based on something similar contained in a book by Harold Jeffreys, is from Richard. This book is dedicated to him.

L'Aquila, Italy Guido Visconti

Acknowledgements

The idea of this book was born from a paper (never published) I wrote with Richard Goody. I carried out from that expanding the content and scope under the constant assistance from Richard. To write a book takes a considerable amount of time from the technical details to obtaining the permissions for the quotations. I have a debt to my student, Paolo Ruggieri, who took time out of his Ph.D. thesis to assist with Latex. Bruno Carli read the final version of the manuscript and made several suggestions to improve the text. My wife, Annamaria, and daughter, Sara, were patient enough to endure the many absences from the daily routine.

Contents

Chapter 1
A Summary of the Problem

Any discussion needs a base of knowledge and this chapter is necessary to summarize a few things that are essential in the debate on the climate change. There is a minimum of simple formulas involved in the introduction of concepts like the greenhouse effect or the climate sensitivity. The global warming is attributed to an intensification of the greenhouse effect which is due basically to gases that were present in the Earth's atmosphere before the industrial revolution. The concentration of some of these gases has been perturbed by the human activity and this is one of the few facts that can be proved experimentally. Different questions arise related to the sensitivity of the climate system (i.e., how many degrees will the temperature rises for a given increase of greenhouse gases, see par. 1.2) and to the mechanisms by which a production of greenhouse gases by natural or anthropogenic processes will impact on their atmospheric concentration. Naturally, we need a definition of climate as distinct from everyday weather. The most simple and efficient distinction between weather and climate is a popular saying "Climate is what you expect, weather is what you get". For some strange reason, the saying is attributed to Mark Twain but for the first time appears in a novel "Time enough for love" by the science fiction writer Heinlein (1973). However, it was explicitly reported by Edward Lorenz in a paper never published (Lorenz 1997). This distinction gives immediately the idea that climate is something of an average of the weather day after day. Each one of us (even without a degree in meteorology) can describe what is the climate in his city making unconsciously an average based on his experience that describes the characteristics of the winter or summer. This qualitative attitude toward the climate is one of the problems to be discussed several times in this book because it is apparently conserved at the highest scientific levels.

G. Visconti, *Problems, Philosophy and Politics of Climate Science*,
Springer Climate, https://doi.org/10.1007/978-3-319-65669-4_1

More recently, especially thanks to some statistician suggestion (Chandler et al. 2010), climate has been defined as "the distribution of weather" and climatic change must be intended as changes in this distribution. They observe that most of the dangerous climate originates from weather extremes. These not only include heat waves but also precipitation extremes (drought and floods). We will discuss several times the different definitions of weather and climate and their implications.

1.1 The Greenhouse Effect

Planet Earth, like all of the other planets of the solar system, receives its energy (radiation) from the sun, contained mainly in the visible part of the spectrum. The surface of the Earth receives on the average about 350 W/m^2 but only part of this power (about 70%) is absorbed by the surface–atmosphere system. The absorbed energy heats the surface and the atmosphere, and these two components reemit energy in the infrared.

Reflection and absorption of solar radiation take place both at the surface and in the atmosphere. In the latter, reflection takes place mainly in the clouds while the main absorber is water vapor. The same thing happens for the emitted infrared radiation absorbed other than from water vapor (chemical formula H_2O) also by other gases present in the atmosphere like carbon dioxide (CO_2), methane (CH_4), and nitrous oxide (N_2O). These are known as greenhouse gases (GHG) and there are a few others. The absorption heats the atmosphere and then the surface and at the equilibrium the total absorbed solar power must be exactly equal to the net power emitted in the infrared. In the absence of such an equilibrium, for example, with the absorbed power less than the emitted power, the Earth would warm forever and vice versa. In the absence of an atmosphere (as the Moon), the emitted radiation would not be captured by the atmosphere and the surface temperature would be about 19 C below zero instead of the comfortable 15 C above zero that we experience on the average today. A difference of 34 C gives a numerical value for what is known as **Greenhouse effect**.

This term is somewhat misleading because it refers to an effect that does not exist. Usually, you are told that a greenhouse works on the same principle, that is, the solar radiation penetrates the greenhouse but the infrared radiation (heat) cannot escape, blocked by the walls made of glass or plastics. Actually, even if you build a greenhouse with transparent walls in the infrared the greenhouse would heat anyway as long as you keep windows and doors closed. As a matter of fact, what is causing the heating is the blocking of the air circulation inside the greenhouse. You can easily demonstrate that by parking your car in the Sun with closed windows.

The equilibrium temperature is a function of the amount of greenhouse gases present in the atmosphere. It is quite surprising that their concentrations are very small. The highest corresponds to water vapor with amount of the order of 10 g/kg of air. The carbon dioxide on the other hand has concentration around 400 molecules for each million of molecules of air (what is called 400 part per million (ppm)) and that is about 0.6 g for each kilogram of air. In conclusion, the atmosphere contains 99.9% gases with no effect whatsoever on climate except for the 0.1%.

If the concentration of a greenhouse gas increases, the absorption of infrared radiation in the atmosphere increases so that at its top there is a deficit in the energy balance (net infrared radiation (IR) is less than net solar radiation). In order to reestablish the equilibrium, the temperature of the surface must increase up to the point that the emitted radiation equals the absorbed one. The power deficit just mentioned is called **radiative forcing** and for a doubling the carbon dioxide concentration is about $4\,W/m^2$.

From 1750 until now, it has been evaluated (only computationally) that the total radiative forcing is due to 64% of carbon dioxide, 18% of methane, and 6% of nitrous oxide. The remaining 12% could be attributed to other greenhouse gases "Computationally" which means that nobody has obtained so far an experimental confirmation of these data. Someone has compared the change in temperature to the water level in a bathtub that receives a constant flow of water from the faucet (solar power). The water level stabilizes when the flow from the faucet equals the flow from the sink (outgoing infrared radiation) which is a function of the water level. If the sink is partly blocked (adding of greenhouse gases), the water level rises until the outgoing flux equals the flux from the faucet.

1.2 The Climate Sensitivity

One of the main problems in climate science is to evaluate how much changes the average temperature of the Earth has for an assigned radiative forcing. The answer to this apparent simple question is called **climate sensitivity** which is defined as the change in temperature (in degrees Kelvin, K) for a forcing of 1 watt per square meter. Again today this parameter is mainly evaluated using models and its "bare" value is around $0.3\ K/wm^{-2}$. Bare means that this would be the sensitivity without any feedback mechanism, that is, for a doubling of the carbon dioxide concentration, the average temperature of the Earth could increase by $0.3 \times 4 = 1.2\,K$ (remember here the $4\ wm^{-2}$ of the previous paragraph). Actually, the IPCC document gives a much larger uncertainty, that is, the change in temperature could be between 1.5 and 4.5 K and this could be attributed to the different feedback mechanisms present in the climate system. The value 0.3 corresponds to an atmosphere containing only carbon dioxide. However, just the presence of water vapor introduces an amplification effect. When the concentration of carbon dioxide (or any other greenhouse gas) increases, the temperature of the atmosphere increases with the effect of increasing the saturation pressure for water vapor. More water can then evaporate and increase the mass of this gas in the atmosphere. The net result is that the initial increase in temperature corresponds an increase in concentration of a very efficient greenhouse gas like water vapor that causes a further increase in temperature. This mechanism is known as **water vapor feedback**. There are several other feedbacks based on clouds or the Earth's albedo and they can be positive (in the same direction of the initial cause) or negative (in the opposite direction). There is a very simple expression which relates the climate sensitivity to the **feedback factor**, f. If a perturbation ΔF in the radiative flux is introduced, this will produce a temperature change ΔT given by $\Delta T_0 + f\Delta T$. The first term is simply the definition of climate sensitivity $\Delta T_0 = \lambda_0 \Delta F$, while the second term is the effect of the feedback that we assume to be proportional to the temperature through the feedback factor f. We have then $\Delta T = \Delta T_0 + f\Delta T$ from which

$$\Delta T = \frac{\Delta T_0}{(1 - f)}.$$

From this simple expression, we see that with a feedback factor equal to 0.5 the temperature change doubles and tends to infinity for a feedback factor reaching 1. This explains why for a carbon dioxide doubling we may reach 4.5 C starting from a $\Delta T_0 = 1.2$°C.

The introduction of the feedback factor complicates the attribution of the causes for a generic increase in temperature. Besides, it can be easily shown that an uncertainty in the feedback factor introduces a much larger uncertainty in the sensitivity. It is worth to notice that the water vapor feedback has been measured many years ago in a very simple situation above the ocean. It seems to go in the direction we just described. This is one of the few experimental results that can be counted in the climate science. As an example, the climate sensitivity has never been measured experimentally. Although we will see later, many suggestions have been made.

1.3 The Global Warming potential (GWP)

Not all the greenhouse gases we mentioned have the same effects on the climate and then may be useful to invent a classification that may help in deciding about strategies to limit their production: the most dangerous first and those relatively unharmful last. This classification may depend on a number of parameters that range from the efficiency it absorbs infrared radiation, to its concentration, and to the modalities of its atmospheric release. All these data must be condensed in just one, that is, the **global warming potential (GWP)**. We consider one kilogram of gas and calculate its contribution to the radiative forcing during a certain period of time. The same thing could be done for a reference gas (usually carbon dioxide) and the ratio between these two quantities is the GWP. It is clear that for CO_2 GWP = 1. A very important factor which determines the GWP is the **lifetime** of the gas and this indicates that time it takes for the molecules of that gas to disappear from the atmosphere. This time ranges from 100 years for carbon dioxide and nitrous oxide to 10 years for methane to 3000 years for sulfur hexaflouride. This lifetime also indicates roughly the interval of time for which the impact of that particular gas is important. So methane has its maximum effect in about 20 years after its release, the nitrous oxide 100 years after its release, and so on. For the indicated times, the GWP ranges from 60 for methane to 300 for nitrous oxide. This means that 1 kg of methane produces an effect equivalent to 60 times that of 1 kg of carbon dioxide. Luckily, the amount of methane present in the atmosphere is about 200 times less that of CO_2 and then the net effect is much smaller at least for the present concentration. Again all these numbers are just theoretical numbers and there is not an experimental measurement of the GWP.

1.4 The Carbon Cycle: How the Concentrations of CO_2 and CH_4 are Determined

In 1958, Charles Keeling of the Scripps Institute of Oceanography began to measure systematically the concentration of carbon dioxide in the Earth's atmosphere from a station at 3000 m altitude at Mauna Loa, Hawaii. In a few years, Keeling noticed that the concentration of this gas was increasing and that superimposed to the trend there was a seasonal cycle with maximum concentration in the winter season and a minimum in spring–summer in the northern hemisphere. The concentration measured by Keeling in 1958 was 353 ppm and after 57 years has reached 400 ppm. The seasonal cycle can be explained because during the spring–summer season, plants take up CO_2 for the photosynthesis and most of this activity takes place in northern hemisphere continents. The increase of the carbon dioxide concentration with time is now attributed to the fossil fuel consumption but what are the proofs for this explanation? The proof has been found almost at the same time of the Keeling measurements but then risked to be corrupted by the human intervention. The idea came from the Swiss geologist Hans Suess based on the fact that carbon in nature appears under three isotopes: the most abundant (99%) is the so-called Carbon 12 (^{12}C), the other stable isotope is the ^{13}C, and then there is the unstable and radioactive ^{14}C. This is well known because the carbon dating method is based on it. The last isotope originates from the interaction of the atmospheric nitrogen with the cosmic rays. As all the other isotopes the chemical properties are the same so that ^{14}C is assimilated by plants and animals as the stable carbon. As a consequence, it also contained in those plants buried for several hundreds of thousands of years from which fossil fuels probably originate. During this long period, outside the influence of cosmic rays, the isotope decays again as nitrogen so that the fossil fuels are impoverished of ^{14}C with respect to the atmospheric composition. When the fossil fuels are burned the carbon dioxide coming from them has a lower proportion of ^{14}C, so that the more fossil fuels burned the less, which is the proportion of ^{14}C in the atmosphere. There is a similar effect (much more complex) related to the photosynthesis that involves the ^{13}C. The atmospheric nuclear test at the end of the 50s and the beginning of the 60s risked to destroy such precious toy. However, some genial trick has permitted to establish that the increase of the carbon dioxide in the atmosphere is definitively due to the combustion of fossil fuels. In this case (a drastic difference from what we will see later) the proofs are experimental. The son of Charles Keeling (Ralph) was able to measure the decrease in the concentration of atmospheric oxygen due to the combustion of fossil fuels and its seasonal cycle which is the opposite of the carbon dioxide. It is worth notice that the concentration of oxygen is more than 500 times that of carbon dioxide, so that the measure is quite complex (see for example Weart 2008 for a detailed history).

The concentrations of carbon dioxide and methane are determined by a balance between sources and sinks. When these two data are known, it is relatively simple to evaluate the gas concentration. The problem again is that these two data are subject to some political influence. The main source of the carbon dioxide is the production of energy and so each single country should have an inventory of such sources. Another important source is the forest management (deforestation or afforestation) or in general what is indicated as LUC (land use change). In both cases, it can be imagined the degree of uncertainty present in the data. The latest data refer to the year 2014 and for that year the global amount of CO_2 produced is around 36 billion tons. This number refers only to combustion of fossil fuels and cement production and does not include LUC that may contribute another 10%. The top producer is China with 10.5 billion followed by United States (5.3), Europe (3.4), and so on. The pro-capita consumption groups together countries with quite different populations like USA (16.5 tons) and Saudi Arabia (16.8) with China trailing behind (7.6).

Of the total amount of CO_2 produced about 47% ends up in the atmosphere, 27% is taken up by vegetation, and 26% is absorbed by the oceans. About half of the CO_2 produced is then destroyed so that if we should stabilize the concentration in the atmosphere, production should decrease drastically. Stabilization could be obtained for example by stopping the positive growth (today around 2% per year) and going to a negative growth of 10% per year. This perspective is quite unrealistic considering that most of developing countries like China, Brazil, and India are far away from western countries in terms of pro-capita consumption (or production).

We like to finish with some provocative argument. First of all, how about the natural sources of CO_2 like the volcanoes. The estimates available today give a total contribution of about 1 gigaton that is quite negligible. We have also mentioned that one of the main sinks for CO_2 is the vegetation and this brings us to talk about deforestation and its role in the oxygen production. The two processes we have mentioned, photosynthesis and respiration and decay, cannot control the amount of oxygen in the atmosphere on short-term basis. As a matter of fact, it is estimated that the total amount of carbon present in the atmosphere (as CO_2) is about 800 gigatons so that the annual contribution from fossil fuel combustion is about 1%. Even if one imagines to oxidize all this carbon, the oxygen would decrease by 0.2%. Even if all the carbon present in the biosphere would be oxidized the oxygen would decrease by 1%. The regulation of the oxygen present in the atmosphere does not depend on the presence of large forest (i.e., Amazon) but it is a much more complex process and involves mechanisms related to the creation of the new Earth crust and more in general to plate tectonics.

This brings us to clarify one last point, the so-called spontaneous regulation of the carbon dioxide concentration. There are recurrent voices about the natural self-regulatory mechanism especially from "creationist" circles. According to this theory if the CO_2 increase so does the temperature and the rain with it. The rainwater will accelerate the erosive processes and so the carbon dioxide will decrease because it will combine with calcium and magnesium silicates to form carbonate and silicate products. These are transported by rivers in the ocean where they form the sediments that are recycled to the surface by metamorphic and volcanic processes. This mechanism will regulate the carbon dioxide and reduce the greenhouse effect. The problem is that the time frame for the cycle to be completed is about 100 million years, a period long enough to destroy several technological civilizations. This cycle can be reconstructed for the last half billion years that brought about the reduction of the concentration of carbon dioxide by a factor 25–30 respect to the present level. Unfortunately, the cycle of the fossil fuels consumption acts on a time frame of the order of 100 years and is comparable to the present life expectancy. We do not expect any miracle after all.

The Scientific Basis

A climate scientist and a humanist at the end of each chapter will discuss what we intended to say and what they have understood. The humanist is not moved too much by the rhetoric of science and keep pressing the climate scientist. This entertaining interval was suggested by Richard Goody which possibly got the idea from the debate between a bothanist and a logician in the Harold Jeffrey's book *Scientific Inference* (Jeffreys 1973). A somewhat similar discussion is contained in the Eugene Ionesco play *The Rhinoceros* (Ionesco 2007)
H: You have tried to make a crash course on the "truths" we know about global warming but I see that except for some data most of those truths are just hypothesis.
CS: One of the most singular aspects of the climate change studies is that the data are really scarce and this is one of the reasons we will debate. It is not clear that climate science could still be considered as a classical scientific endeavour, that is, if it follows the chain: experimental facts—falsifiable theory—prediction of new "experimental facts". We will see that in most cases this is not possible and we should find some new perspective.
H: In particular, I understand that most of the radiative effects or forcing of the greenhouse gases are entirely based on theoretical considerations.
CS: The physics and chemistry of the greenhouse gases is quite clear and the spectroscopic data are the basis for many of the calculations reported. However, direct measurements for example of the radiative forcing are lacking as well as experimental evaluation of the climate sensitivity. Some of these data could be obtained through satellite measurements but there are difficulties of various kinds.
H: I understand that even in this case there are groups that count more than others. It looks like there is some conflict between meteorologists and climate scientists. The tragicomic aspect is that the expected cuts in the US on these researches will affect both because the politicians do not know the difference.
CS: Some spaceborne measurements may solve experimental problems related to the greenhouse effect. As an example the problem of the energy budget of the Earth–atmosphere system has received a very important contribution from these measurements. However, we must keep in mind that in the case of the greenhouse effect we are talking about very small signals. For example, if the radiative forcing due to a doubling of CO_2 is of the order of 4 W/m^2 and this happens in 50–100 years time you need to measure changes that ranges between 40 and 80 milliwatt per square meter per year and these are really small variations with respect for example to the natural variability.
H: How this problem can be solved keeping in mind that this should happen in a time frame compatible with the government decisions.
CS: This is one of the aspects we will try to illustrate using not necessarily direct measurements but strategies which involve the credibility of the models.

H: The estimates of the climate sensitivity are very much uncertain. How can you accept a change of temperature between 1.5 and 4.5 for the most plausible scenario? If the principles on which the models are based are the same, how can you justify such a range?
CS: The absolute value may be a minor problem as long as the effect is solid. However, the necessity to inquire why models based on the same principles give such different results is real. This may be related to the fact that construction of models is not strictly a scientific but it is more like an engineering endeavour with a lot of patches. Sometimes, the same model gives very different results by changing a single parameterization.
H: The main progresses have been made in the comprehension of the cycle of the different greenhouse gases. I understand this is a sector which involves not only physics and chemistry concepts but also the physiology of plants and animals.
CS: These kinds of studies are performed in a vast interdisciplinary sector known as biogeochemical cycles. This sector had an impressive development in the last 20 years with a very sophisticated models and through dedicated field campaign and global measurement network. Recently, a satellite has been orbited to study the carbon cycle. At the present time, study of the biogeochemical cycles may be in a better scientific position than the climatic effects.
H: Sometimes, you got the impression that some published paper has the only purpose to add material to the author curriculum and the impression remain of a very poor scientific approach to the problem. On the other hand, in a recent paper by "guru" Palmer (2016), he encourages young scientist not to engage in the development of ab-initio models but rather to follow a seamless approach to relate weather and climate predictions.
CS: In many scientific sectors, most of the published material could be neglected. In this particular case, this behavior may be dangerous because many of these "useless" papers are used by the so-called negationists that deny the existence of global warming. Some years ago there was a paper published in one of the top scientific journals that tried to demonstrate how models were overestimating the change in precipitation. Today, many experts of that sector claim that the paper should have never appeared on a journal because of neglecting some very important data.
H: There is then the question of the IPCC. How science can be based on the consensus of groups of scientists that are nominated by their government, that works with money from the same government, and so on. Is not this a gigantic conflict of interest?
CS: This is another aspect to be discussed. However, IPCC does not make science because in theory makes a huge review of whatever is published. In practice, IPCC then includes also paper still in the review process or in printing. Theoretically, IPCC should not have a political position and in reality IPCC makes a review of papers which goes all in the same direction.

References

Chandler, R., Rougier, J., & Collins, M. (2010). Climate change: Making certain what the uncertainties are. *Significance*, *7*, 9.
Heinlein, R. A. (1973). *Time enough for love*, Ace.
Ionesco, E. (2007). *Rhinoceros*, Faber & Faber.
Jeffreys, H. (1973). *Scientific inference*. Cambridge: Cambridge University Press.
Lorenz, E. (1997). *Climate is what you expect*. (unpublished manuscript).
Palmer, T. N. (2016). A personal perspective on modelling the. *Proceedings of the Royal Society A*, *472*, 20150772.
Weart, S. R. (2008). *The discovery of global warming: Revised and expanded edition*. Harvard University Press.

Chapter 2
How Climate Is Studied

The study of climate includes a large number of methods and techniques. Basically following the definition given previously, the most elementary method to classify climate is the collection of data on simple variables like temperature, rainfall, and so on. The requirements that climate should be based on several years averages already gives problems related mainly to the quality of the observations. How honestly could one compare a temperature data taken today with the same data taken one hundred years ago? Also some data were just not available in historical time. For example, think about the vertical profiles of temperature.

From this point of view, climate sciences already have some peculiar problems. They apparently utilize the same methods used in other physical sciences with quality degrading. As we go further back in time, another problem arises from the geographical distribution where the data are taken. Considering the technological progress in data gathering today, there are dense geographical networks that collect data, while in the past they were quite sparse. This may create some bias in evaluating the regional climates and their evolution.

Finally, an aspect is related to the distinction between weather and climate. Most of the climate data are actually derived from weather data which at the least in early times had different requirements. Today, a large part of climatic data is still derived from meteorological networks and the work to make them suitable for climate use is considerable.

In the following, we will give a summary of the most used methods to study climate, which remain entangled with the history of meteorology. At the end, a short paragraph is given on how climate is predicted and of course this aspect will be a common feature of all the following chapters.

G. Visconti, *Problems, Philosophy and Politics of Climate Science*,
Springer Climate, https://doi.org/10.1007/978-3-319-65669-4_2

2.1 Meteorology and Climate

At times we have mentioned the weather as the climatic "noise". It is now time to mention some recent development of this concept. Even referring to the popular definition attributed to Mark Twain climate is a long-term average of the weather. This simple concept seems to be confirmed by the fact that both the numerical weather forecast and the climate predictions are made basically with the same General Circulation models (GCMs). In the first case, the model is run using initial conditions (like temperature, pressure, etc.) while in the second case, the boundary conditions are fixed (amount of greenhouse gases, solar radiation, etc.). In this case what results from the first "is what you get", while in the second case "is what you expect". We will resume this discussion in the next chapters.

This apparently is what the good sense dictates but it is not based on an objective analysis of the real data. Recently, Shaun Lovejoy of the McGill University has found (Lovejoy 2013), analyzing real data, that there are three regimes for the variables of the weather and climate. Between the weather, with a characteristic period of 10 days, and the climate, with periods longer than 30 years, there is the so-called macroweather. In very simple term if one draws the variations of temperature with a resolution of one hour, find that the maximum fluctuation is around 5°. If the same data are plotted with a resolution of 20 days, the maximum fluctuation is reduced to 1.5° and this value goes back to 5° for a resolution of 100 years. For example, with this resolution in a 100.000-year interval we can observe the ice age cycles with maximum fluctuations of 5°. Paradoxically on very long timescale, climate behaves as the weather (See Fig. 2.1). This discovery could change completely the perspective of climate predictions because (according to Lovejoy) GCMs do not predict the future climate but rather predict the macroweather. This is confirmed by the fact that the fluctuations predicted by the models are too small that may indicate that GCMs are missing some mechanism (connected probably to some slow process like the ocean circulation) that do not make them able to predict the climate but rather a long-term extension of the weather. This characteristic of the GCMs was already evident a few years ago in the analysis of satellite data performed by the group of Richard Goody at Harvard (Huang et al. 2002). This analysis did show some rigidity of the model with respect to the natural variability. This result could have some influence on the possibility to evaluate the quality of the models by comparing the spectra of their fluctuations with the spectra of the natural fluctuations and those produced by the presence of the radiative forcing due to the greenhouse gases.

In the previously mentioned paper by Palmer (2016) the author introduces what he calls the climatic Turing test. Originally, it was formulated to distinguish between human and artificial intelligence. The proposed test was to ask questions to the two entities and if the answers were same, there is no way to tell the difference. In the climatic case, the two entities are modeling and the data. If one compares the output of a high-resolution forecast model after 1-day integration, there is no way to distinguish between model and the real data while it is quite easy to tell the difference between a climate model prediction and the real data. This different

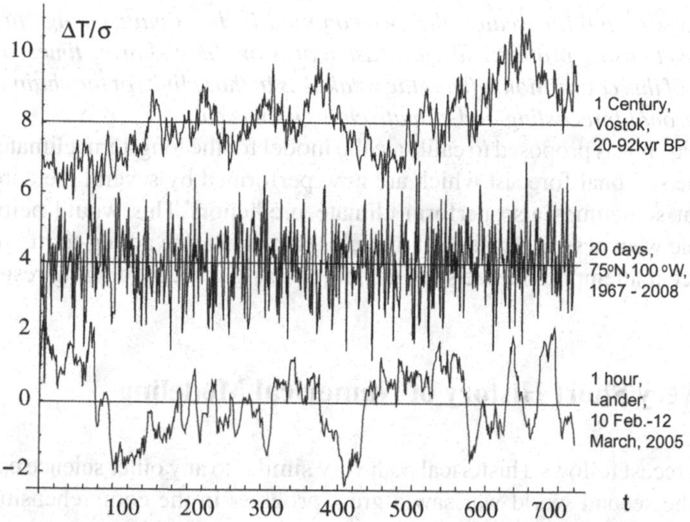

Fig. 2.1 Dynamics and types of scaling variability: a visual intercomparison displaying representative temperature series from weather, macroweather, and climate. To make the comparison as fair as possible, in each case, the sample is 720 points long and each series has its mean removed and is normalized by its standard deviation (4.49 K, 2.59 K, 1.39 K, respectively), the two upper series have been displaced in the vertical by four units for clarity. The resolutions are 1 h, 20 days, and 1 century, respectively, the data are from a weather station in Lander Wyoming, the twentieth-century reanalysis, and the Vostok Antarctic station, respectively. Note the similarity between the type of variability in the weather and climate regimes. Adapted from Lovejoy (2013)

behavior is another important difference between weather and climate that we will discuss often but for the moment we may take Tim Palmer comment:

I think the climate community worldwide does not give enough priority to developing a model which passes the climatic Turing test. We make do with our imperfect models, typically subtracting out the systematic errors against observations when estimating the impact of climate change. Journal referees recognize that this is the best we can do given the current generation of models, and so scientific careers can flourish without having to address the more fundamental question:Why is it so easy to tell the difference between model output and the real world?

Some years ago another possible interaction was proposed between meteorology and climate or between weather forecast and climate prediction. In order to improve the reliability of the latter, Palmer et al. (2008) proposed the so-called "seamless" prediction:

This is where the notion of seamless prediction can play a key role. It will be decades before climate- change projections can be fully verified. However, our basic premise, illustrated by the schematic in Fig. 2.1, is that there are fundamental physical processes in common to both seasonal forecast and climate change timescales. If essentially the same ensemble forecasting system can be validated probabilistically on timescales where validation data exist, that is, on daily, seasonal, and (to some extent) decadal timescales, then we can modify the climate change probabilities objectively using probabilistic forecast scores on these shorter timescales. The magnitude of this calibration reflects the weakness in those links of the chain common to both seasonal forecasting and climate change timescales.

In practice it was proposed to calibrate the model for the long-term climate projections on the seasonal forecast which are now performed by several meteorological centers that sometimes also perform climate prediction. This would help to isolate possible weakness in the several modules that constitute a long-term prediction model. Recent attempt to utilize such technique has given encouraging results.

2.2 A Very Short History of Numerical Modeling

Weather forecast follows a historical path very similar to any other science especially physics. The second world war saw a great progress in the comprehension of the general circulation of the atmosphere with the major contributions coming from Jule Charney, Eric Eady, Hans Ertel, and Carl Rossby. The general problem corresponds to that of a rotating fluid like the Earth's atmosphere and its complexity is acknowledged especially for its mathematical aspects. Jule Charney to solve one of the central problems (the baroclinic instability) had to recur to some sophisticated mathematical function (Charney 1947).

The problem is that the time evolution of the variables which describe the system (like temperature, wind, humidity, etc.) is established through a number of nonlinear mathematical equations. This means that the response of the variables is not proportional to the causes that determine their changes. Or else the instantaneous change of one of the variables depends on a combination of the other (like a product). Charney to solve these equations utilized very sophisticated mathematical methods that were time-consuming because most of the time they required tabulated function. John von Neumann developed integration techniques that could be used on a computer machine (Smagorinsky 1983; Nebeker 1995; Dyson 2012). Toward the end of the 40s, the development of the first numerical computer coupled with these integration techniques brought the first numerical forecast in 1948 based on the theoretical concept of potential vorticity. Progresses are constant and the next big step comes in the 60s with the introduction of the weather satellites that fill the gap to have data on a global basis especially for those regions inaccessible to direct measurements.

In the same years, Edward Lorenz (1963) working on a set of simplified equations to study convention in fluid dynamics discovered that nonlinear systems behave in a very strange way. Notably, the prediction of the motions depends critically on the initial conditions. The solutions were so sensitive to the initial conditions that their precision would be crucial: it was the discovery of the deterministic chaos. Lorenz was a meteorologist who had passed all the stages of the trade, thanks to his genius; today, meteorology is a respectable science that has opened new horizons also to another realm of physics.

The consequences of the Lorenz discovery impact directly on the weather forecasts because the inescapable error in the initial conditions propagates rapidly in the course of the integration and after sometime (3–5 days), the error is so large that the forecast loses any meaning. This corresponds to affirm that the system loses memory of its initial conditions. The situation may be more complex than that because sometimes the precision of the measurement could be acceptable but the resolution of the model is too poor to accept that data. For example, the resolution of some satellite data today is of the order of 1 Km, but many of the forecast models and none of the climate models reach that resolution.

Meteorology is then a very respectable science because it is based on a continuous exchange between the observations (ground based and satellite data) and the numerical methods that have introduced new concepts in science like the assimilation techniques. This is a process that relaxes the solution toward the observed data.

This implies organization and economic efforts that put meteorology in the category of the big sciences. Think about the constant maintenance of the large technological apparatus (radars and remote sensing instruments). To the replacement of more and more sophisticated satellites, think about the huge amount of data produced that have to be processed. This effort requires an aggressive scientific community made by high priests who seat in the forecast centers and peones who are the interface with the community (i.e., weather men). This community has become an important power group that runs heavy international organisms (like the World Meteorological Organization, WMO) and has the tendency to occupy also the climate realm. This community is rather peculiar because it on one hand makes science (scientists are those who make science as Lewontin and Putnam affirm) while on the other has an unmatched applicative side. It is even different from medicine whose scientific basis must be found in other sciences like chemistry or biology. Meteorology has built his own scientific corner which is largely unrecognized. None of its great pioneers (we listed a few but many others could be added) has received the Nobel Prize (for those who care).

2.3 Earth's Climatic History: A Source of Data

At this point one would think that climate can be studied in the same way as meteorology, only on much longer timescales. A scientific study, however, requires the documentation about the relevant data, theories that explain the data, and the same theories (if correct) should reveal or predict new phenomena. Let us see if the study of climate responds to such requirements.

Starting from the data we know that the history of climate on Earth begins with the origin of the planet and that already force us to give up part of the story because the data of what could be the climate a few billions of years ago are rather confused. Any textbook you read on the past climate still give you the division based on different timescales from billions to million to hundreds of thousands years. Actually, we have only a very vague idea of what could be the climate from the origin until half billion years ago and in any case data for that period cannot be used to test any theory. For example, during the last 25 years a new theory has emerged about snowball Earth. This theory argues (Hoffman and Schrag 2002) that for some still obscure reason, the Earth plunged into a state of being completely covered by ice. Geological data seem to confirm this episode that ends about 600 million years ago. However, the climatological data are of such poor quality that does not allow to test global climate models.

Another catastrophic episode of the Earth's climatic history relates to the dinosaur extinction about 65 million years ago. At the beginning of the 80s, a group of scientists that later became famous as TTAPS (Turco, Toon, Ackerman, Pollack and Sagan) proposed in a *Science* (Turco et al. 1983) paper that a global nuclear conflict could have triggered fires of such an extent to fill the atmosphere with soot that could have obscured the Sun causing what was called a "nuclear winter". As a bad example of being carried away, the same TTAPS had proposed (Pollack et al. 1983) that a similar mechanism could have caused the extinction of the dinosaurs. In this case, the cause for the attenuation of light would have been the asteroid that hit the Earth according to the hypothesis formulated by Luis Alvarez some years before. The impact would have filled the atmosphere with dust. Actually, TTAPS had worked already on transient climatic changes due to the volcanic eruption (see for example Toon et al. 1982). In this case, they used very simple models that today any student may use as an excercise. However, considering the political climate of that time and the charisma of some of the TTAPS people, the noise was completely justified. The conclusions remained in a vague area good for *Discovery Channel*, but still that was not hard science.

As we approach more recent times the fog on the climatic data starts to thin out. Typical examples are the ice ages for which there are data that go beyond 4 million years ago. The ice ages are episodes of cooling of the Earth's climate that recurs with typical periods of 20.000, 40.000, and 100.000 years. These are the same periods that characterize the variations of some orbital parameters of the Earth, like the precession of the axis of rotation, its inclination, and the eccentricity of the orbit.

The change in eccentricity is such that the orbit around the Sun goes from a perfect circle to a very slight ellipse. In this way, the Sun–Earth distance changes from a constant value to a variable one. In the present epoch, the Earth reaches its minimum distance from the Sun around mid-winter (winter in the northern hemisphere) and its maximum distance in the summer. It is clear that a modulation of the distance introduces a modulation in amount of the instant radiation received by our planet. The rotational axis of the Earth is not perpendicular to the plane that contains the orbit of the Earth (ecliptic) and makes an angle of about 23.5°. This inclination is responsible for the seasons on our planet: when the axis points toward the Sun, we have summer in that hemisphere, and when it points away it is winter. The inclination is not constant but changes between 22.1 and 24.5° with a period of about 41.000 years. During the periods of maximum inclination, the difference in insolation between summer and winter are larger than during periods of low inclination.

Finally, the rotational axis not only "oscillates" but also "precesses", that is, has a motion similar to a spinning top with a period of about 22.000 years. As a consequence, the position of the seasons around the Earth's orbit changes with time and this corresponds to the so-called precession of the equinoxes. The equinox is the day of the year when night and day have the same duration and this happens at the beginning of spring and fall, while Solstice is the day when the night has the shortest duration (summer solstice). In the present epoch, the winter Solstice (longest night of the year) in our hemisphere happens roughly on December 21, while the minimum distance from the Sun is reached on January 3. The precession of the equinoxes makes this distance to grow constantly so that the opposite situation (winter at the maximum distance) will be reached in about 11.000 years. At this point, it should be clear that the coupling between the precession of the equinoxes and the changes in the eccentricity is the one that produces the largest variations in the quantity of solar radiation reaching the Earth. Without doubt, the changes in the solar radiation related to the orbital parameters have an influence on the ice ages but many problems remain. One particularly important is the fact that while until 800.000 years ago the dominant period of the ice age was the same as the change of inclination (40.000 years); in most recent time, the dominant period is 100.000 years.

There are experimental facts about the ice ages that need explanations especially about the termination phase. As an example in the last ice age, the deglaciation phase was suddenly interrupted by a few episodes of climatic hiccups when the gradual warming was stopped and the climate went back to a glacial one. The most famous episode corresponds to the so-called Younger Dryas about 8000 years ago which had a global character. Further, there is the problem related to the atmospheric composition. At the onset of a glaciation, the decrease in temperature is accompanied by the decrease in the concentration of carbon dioxide and methane and by the increase in the atmospheric aerosols. All these phenomena do not have precise quantitative data but remain as "facts" that are explained in part or completely by ad hoc theories. Studies ran from purely phenomenological to rather sophisticated models and their results seem to point out an important role for the ocean. An excellent review on the ice ages can be found in Rapp (2010).

2.4 The Oceanic Circulation

The role of the oceans in the determination of the climate has received a wide attention by the scientific community. Two examples may give an idea how heated is the debate. One refers the Gulf stream and the other the so-called thermohaline circulation.

The Gulf stream is a surface current that moves from the tropical Atlantic ocean toward higher latitudes along the East coast of North America. For many years, people have assumed that such a current would explain the temperature difference observable between the East Coast regions of North America and the those of Northern Europe. The former are about 5 C colder than the latter. In 2002, a group of American scientists headed by David Battisti and Richard Seager (Seager et al. 2002) showed how the effect of the oceanic currents was negligible with respect to the contribution of the atmospheric currents. A possible proof is the existence of a similar oceanic current in the Pacific known as Kuroshio Current that originates from the coasts of Japan and then moves toward northeast but misses the coasts of Canada and USA. On the other hand, winters in regions like Vancouver or Seattle have very mild winters that cannot be attributed to this oceanic current. The importance of the atmospheric currents is confirmed by the fact that the maximum heat transported by the ocean at about 20° North amounts to two thousands billion watts, while the atmospheric contribution is about six thousand billion watt at 40° North. This discovery made by Battisti and Seager has been almost completely ignored by those who sustain the role of the Gulf stream. Actually, it has a great importance for the climate debate because at the present it is thought that the mechanism for the ice ages depends on the intensity of the Gulf stream.

As a matter of fact, the Gulf stream is the observational part of more complex phenomena known as thermohaline circulation. Surface water in the Gulf of Mexico work as a giant pot where the evaporation is much larger than precipitation of freshwater. As a consequence, these warm and salty water generate the Gulf stream that during their journey toward higher latitudes become more salty and cools and so are "heavy" with respect to the surrounding water. This causes the water to sink near the polar Arctic generating what it is known as North Atlantic Deep Water (NADW). According to the accepted theories, this deep current is part of global circulation that makes the deep water to emerge in different points of the world ocean like the Antarctic. The deep current is also known as Meridional Overturning Circulation (MOC) and has been popularized as the Great Oceanic Conveyor by its inventor Wallace Broecker. If the conveyor is "on" heat is transported from the tropical regions to the Arctic Circle and the planet is an interglacial (like today). If the conveyor is "off" the high latitude region in the Atlantic cool off and the planet is in an ice age. A way to turn off the conveyor is to dilute the high latitude salty water through the melting of the ice sheets. This theory is so popular that even Hollywood believes it as in the movie *The day after tomorrow*.

One of the main critics of this Hollywood theory is Carl Wunsch, a MIT oceanographer. Among the softest things he has written is that the thermohaline current can be observable only at the surface so that to utilize this concept for various theories is like pretend a meteorologist to use only a small region of the atmosphere for his forecasts. Actually, the Gulf stream is a consequence of what in oceanography theory is known as intensification of the western boundary current. In the Atlantic, the dominant wind system is such that at the tropics the wind forces the water toward the west (trade winds) while at middle latitudes, the water is forced toward the east (westerlies winds). In this way, some kind of vortex is generated in the Atlantic basin with a current adjacent to the North American Coast. For a number of reasons, this part of the current tends to be intensified, which is the origin of the Gulf stream. In order to exists the Gulf stream that needs the winds, a concept reaffirmed by Wunsch in a letter to *Nature* in 2004 (Wunsch 2004)

Sir—Your News story "Gulf stream probed for early warnings of system failure" (Nature 427, 769; 2004) discusses what the climate in the south of England would be like "without the Gulf stream". Sadly, this phrase has been seen far too often, usually in newspapers concerned with the unlikely possibility of a new ice age in Britain triggered by the loss of the Gulf stream. European readers should be reassured that the Gulf stream's existence is a consequence of the large-scale wind system over the North Atlantic ocean, and of the nature of fluid motion on a rotating planet. The only way to produce an ocean circulation without a Gulf stream is either to turn off the wind system, or to stop the Earth's rotation, or both. Real questions exist about conceivable changes in the ocean circulation and its climate consequences. However, such discussions are not helped by hyperbole and alarmism. The occurrence of a climate state without the Gulf stream any time soon—within tens of millions of years—has a probability of little more than zero.

Wunsch is a critic of the conveyor belt. He thinks that the complexity of the oceanic circulation can hardly be reduced to a one-dimensional model that even if it were true, the current would be extremely weak. In his own words (Wunsch 2010)

*Broecker (*1991 *, and many other papers), building on a sketch of Gordon (*1986*), reduced the discussion of the paleocean circulation to that of a one-dimensional ribbon that he called the "great global conveyor". Its rendering in color cartoon form in Natural History magazine has captured the imagination of a generation of scientists and non-technical writers alike. It is a vivid example of the power of a great graphic, having been used in at least two Hollywood films, and has found its way into essentially every existing textbook on climate, including those at a very elementary level. It is thus now a "fact" of oceanography and climate. (Broecker* 1991*, himself originally referred to it as a "logo," and it would be well to retain that label.)*

This brings back the question whether the ice ages could be driven by a massive injection of freshwater at high latitude due for example to melting of the Arctic ice sheets. This was the theory about the Younger Dryas: The melting water gathers to a large interior basin (Lake Agassiz) whose flooding in the north Atlantic stops the conveyor, making the Northern hemisphere to return suddenly to an ice age. However, they discovered (Lowell et al. 2005) a difference of about 1000 years between the Agassiz floods and the beginning of the YD. In a more recent paper appeared in *Nature* (Murton et al. 2010) it is shown that even if that flood discharged it did toward the Arctic rather than in the North Atlantic. Here we have another side of the theories on the climatic change. Most of them are ad hoc theories based on the misuse of doubtful concepts (the conveyor).

The ocean then is fundamental in the determination of climate because it is an element of the system which has a capability of absorbing heat thirty times that of the atmosphere but its slowness does not seem to be compatible with the funding system. Wunsch says (Wunsch 2001):

Perhaps the greatest climate puzzle is whether one can find a way to study its "slow physics" (and chemistry and biology) within funding systems based on year to year budgets, high frequency elections, short tenure deadlines, and the general wish for scientific results in the short term. Understanding the ocean in climate is a surely a multi-decadal problem, at best.

There are then several problems to solve with the past climate and this is quite clear in the words of the inventor of the conveyor, Wallace Broecker (Broecker 2010). He

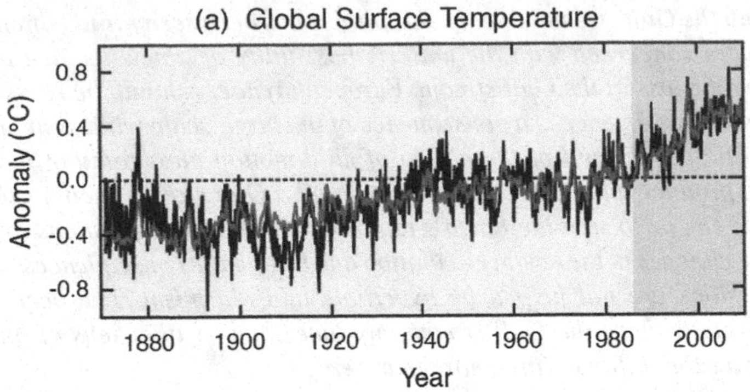

Fig. 2.2 The anomaly in the global average temperature in the last 140 years. The *pink background* indicated the year of the most recent increase in the forcing. The *red line* shows the average while the *black line* includes the fluctuations, Courtesy of IPCC report. Reprinted with permission by IPCC

remarks that his thinking was only marginally influenced by the simulations that have tried to reproduce the paleoclimate. He also observes that until models are unable to simulate the past it will be very hard to believe their predicting of the future.

A quite clear assertion considers that as we have seen the success in reproducing past changes are rather sparse and occasional (ad hoc). Besides the reconstruction of the past climate is still very qualitative and strictly related to the geological research.

2.5 Climatic Changes in Recent Times

Things should change as we approach our time as it is shown for example by the temperature data that now refer to the last 150 years Fig. 2.2. However, even this record is not completely reliable because the temperature standards used 150 years ago were not the same as today. We may affirm that reliable data could be assumed starting at the beginning of 1900. The curve whatever its origin shows a constant increase from 1900 to 1945 and then level off for about 30 years and they resume the increase until the end of last century. After this period there is a new apparent stop in the warming referred as *global warming hiatus* and has produced an avalanche of papers and discussions. The curve presents quite large fluctuations that may be useless to explain. During the past century, we had a number of catastrophic volcanic eruptions that may have influenced the climate. This is because the stratospheric aerosols that forms after the eruptions may reflect the solar radiation and cool the climate. Actually, the most important eruptions are Santa Maria (1902) in Guatemala, Agung (1963) in Indonesia, El Chichon (1983) in Mexico, and Pinatubo (1991) in the Philippines. The problem also in this case is that most of the data are missing except for the last two events. Actually, only the Pinatubo eruption has been properly monitored from the ground and from the satellite. The most important tool to obtain data from the ground is the lidar. In the case of the Agung eruption, Giorgio Fiocco then at the MIT had just invented it and the quality of the data was not enough to prove any kind of theory. We will talk a little longer on this subject considering that of the most popular geoengineering scheme is based on the assumption that volcanic eruption cools the climate which may be not. It is enough to say that in 1978, Jim Hansen and coworkers (Hansen et al. 1978) could claim a complete understanding of the climatic effect of Agung eruption. Today, Canty et al. 2013 and Fujiwara et al. 2015 the volcanic signal is absent from a careful examination of the experimental or reanalysis data.

In spite of these uncertainties, many people have made a career out of them just to confirm that science is made by men. As we mentioned before, the temperature curve has generated ferocious discussions that have interested even the magistrate as in the famous episode of climate gate (see for example *Nature*, **468**, 345, 2010). However, a very qualitative discussion attributes the constant temperature increase up to the second world war to the increase of the greenhouse gases. At the end of the war, the recovery of the industrial activity produces huge quantities of aerosols that may have compensated the warming effect of gases. This continues until the western countries develop legislation to contain the atmospheric pollution. These measures were so efficient that starting from the early 70 s or 80 s the ever increasing production of carbon dioxide determined a new increase in the temperature.

For the last 50 years, we have now a quite precise data known as reanalysis. This process reconstructs 3D meteorological maps than could be used as a test for climate models for some kind of 'hindcast'. It is quite strange that this has been attempted only partially.

2.6 How the Observations Are Made

The experimental methods used to obtain climate data are just an extension of those used in meteorology in the sense that the same database that covers an adequate period of time may be used in climate studies. The real problem arises with the data older than 50–60 years and the question is the intercomparison, that is, the evaluation of a data taken today with the older data even with a simple thermometer. It is not just the absolute value of the temperature but rather its precision. Do not forget that we are talking about changes of 0.5 C in 50 years and so changes of 10 thousandth degrees per year. It is true that the global data result from some kind of mean but the precision of measurement should be high.

We are talking about 50 years because this is the time interval covered by the reanalysis techniques. These have been developed at the end of the 70s and aim to make the measured quantities consistent with the model forecasts. In this way, data produced by models in some grid point are corrected by the observations that were not available at the time of the forecast. On the other hand, models may complete the data in those geographical regions where measurements were not taken. If we only refer to the reanalysis of the European Centre for Medium Range Weather Forecast (ECMWF), the first project ERA (ECMWF Reanalysis) covers the period from 1978 to 1994 (ERA-15). The second project (ERA-40) covers the period 1957–2002 and ERA-Interim goes from 1989 to the present. This is an archive that gives the maximum reliability (see for example www.ecmwf.int/en/research/climate-reanalysis.) However, today people talk about period of time distant hundred of thousands years and pretend they have the same precision and reliability.

We can make the example of the ice ages. In this case, the breakthrough was the discovery of the Urey thermometer. At the beginning of the 40s, Harold Urey had found that the ratio of the most abundant oxygen isotopes (^{16}O and ^{18}O) was a function of the evaporation and condensation processes. This means that the continental ice sheets are enriched in the lighter isotope. The cycle is that of the evaporation at low latitude, when the clouds are enriched of the lighter isotope. The precipitation that forms the ice sheets is then enriched in the ^{16}O isotope. On the other hand, the ocean water are enriched in the heavier isotope. These water provide the oxygen in the calcium carbonate of some marine shells that go to constitute the oceanic sediments. A core drilling of the sediment may reveal an increase of ^{18}O that indicates an increase in the volume of the ice sheets. Cesare Emiliani (the brain drain has an ancient history) in the 50s was the first to show that from a core sample it was possible to obtain an estimate of the ice volume and then of the temperature. This technique has furnished, starting from the 80s a vast amount of data on the variations of the ice sheets volume in the last 3 million years. The spectral analysis of these data has permitted not only to isolate the harmonics at 20.000, 40.000, and 100.000 years (as seen before) but also minor changes before the onset of the ice ages or before their termination. The literature on this is immense (a good reference is Rapp 2010) but the conclusions we may draw on the evolution of the present climate are relatively poor. For example, the simultaneous changes of the carbon dioxide and the temperature are something like the old chicken and egg dilemma. There is no chance that such a data could be used to estimate the climate sensitivity because the environmental conditions were so much different from today. Besides while there is a quite precise estimation of the greenhouse gas concentration, the temperature changes may have a large relative error because their evaluation is rather indirect.

In any case, the isotope method and other similar techniques have contributed to a better knowledge of the Earth system involving aspects that go from geology to biology to the plant physiology. Everything goes in the general direction of a better understanding of the climate system but does not give direct clues on how it will be our near future.

2.7 Satellite Observations

The satellite observations related to the climate studies may date back to the end of the 60 s when NASA orbited TIROS 3. This meteorological satellite operated a Michelson interferometer that is an instrument that measures the infrared radiation coming from the Earth and could make very accurate spectra of this radiation. The name of the instrument was IRIS (Infrared Interferometer Spectrometer) and was invented by Rudolf Hanel of the NASA Goddard Space Flight Center. Since then IRIS (or similar instruments) has been employed in a number of sondes and satellites starting from the Mariner 9 that was able to characterize the Mars atmosphere in 1971 (Hanel et al. 1992). The infrared spectra of the Earth were characterized by the absorption of four gases, water vapor, carbon dioxide, ozone, and methane. The study of such absorption bands allows to obtain quantitative estimates of the interactions between the temperature changes at the surface and the feedback processes involving the different gases.

This technique received an important contribution from a research group at Harvard (Jim Anderson, Richard Goody, Stephen Leroy) that using the spectra of Nimbus 4 was able to show that the natural variability of the temperature was much larger than that predicted by the general circulation models. Based on this idea, but especially on the possibility of calibration in orbit, at the beginning of the new century, the proposal is made for a spaceborne instrument capable of absolute spectral measurements. The satellite that could carry the instrument initially is named Arrhenius in honor of Svante Arrhenius the Swedish chemist that was a pioneer of the greenhouse effect. The iteration that followed the project becomes CLARREO (Climate Absolute Radiance and Reflectivity Observatory) that is accepted by NASA but then is somewhat frozen in 2012. This proposed experiment could be the only one capable to measure quantitatively the evolution of the greenhouse effect on Earth, and so gives a definitive answer of what could be the anthropogenic contribution to the perturbation of the energy budget of the planet.

The other instrument on the satellites of the CLARREO project is a GPS receiver. We are now familiar with the GPS but most people (especially users) do not know that an important correction to the GPS signal is due to the atmospheric water vapor. This is because the refractive index of air depends on the quantity of water vapor that is present. This means that the GPS data can be utilized to obtain both the water vapor and the temperature of the atmosphere. These data are of such importance that is utilized to improve the weather forecast. Even today there are immense quantities of data referred to the temperature profiles of the Earth's atmosphere that covers the last twenty-five years. CLARREO utilizes this technique with the additional bonus of using calibrated data on international standards. The final product would be the evolution of the thermal structure of the atmosphere determined with a precision never reached before. These measurements can be carried out for a very long time, because they are calibrated and can be easily compared and could bring back the global warming in the realm of the physical problems. As we may have already mentioned that CLARREO was frozen by NASA, it has been canceled definitively by the Trump administration.

2.8 The Climate Through its Fluctuations

In 1975, a researcher at NCAR, Cecil Leith, published a paper on the *Journal of Atmospheric Science* (Leith 1975). In this work, Leith argued that the could be treated as a thermodynamic system (like a gas) where some quantities (like temperature) may exhibit spontaneous variations even in the absence of external forcing. In a gas, this is due to the fact that the molecules are subject to a random motion that is responsible for the fluctuations. Twenty-three years before Herbert Callen and Richard Green of the University of Pennsylvania had shown (Callen and Greene 1952 thermodymanic system forced out of the equilibrium would return to the normal condition in a characteristic time that was the same of the decay time of the spontaneous fluctuations. Leith's idea was based on this hypothesis but was applied to the climate system. The climate system has an intrinsic "noise" that for example could be identified with the weather and may produce the observed fluctuations. In studying these fluctuations we may assume that their decay happens in a time which is the same of the decay time of the whole climate system. We can imagine a very simple climate system reduced to a forcing like the solar component and a damping like the infrared radiation. This could be assumed proportional to the temperature: if the solar radiation increases the temperature, more infrared radiation is emitted and the system goes back to equilibrium. The rate of decay depends on the thermal capacity of the system and the climate sensitivity and actually is the product of these two quantities. In a system with low climate sensitivity the return to the equilibrium is faster than in a system with high climate sensitivity.

In practice, the measure of the climate sensitivity is reduced to a measure of this decay time and this can be done by studying the so-called autocorrelation of a climate variable. From the autocorrelation, it is possible to determine the decay time and once the thermal capacity is known the climate sensitivity can be determined. A similar procedure has been used with the 150 years temperature records and the resulting climate sensitivity is just too low (Kirk-Davidoff 2009). From the discussions that followed, it seems to be clear that the idea may be correct while the data are not adequate. However, Leith idea could also be applied to the autocorrelation of the emission spectra but again this requires a calibration of the measured spectra. A more sophisticated analysis shows that with a similar technique it is possible to compare the natural variability with that predicted by the models. We already mentioned this point and we will discuss it again later. Again the method outlined the stresses the occasion missed with the CLARREO project.

2.9 How Climate Predictions Are Made

Before discussing this aspect, it may be useful to introduce even another definition of climate. Jonathan Rougier of the University of Bristol and his coworkers also challenge the idea that climate is "average weather" (Stephenson et al. 2012). They claim that weather is a measurable aspect of our ambient atmosphere, notably temperature, precipitation, and wind speed. Climate as consequence is a subjective distribution for weather and it is not "average weather" because this would be just a summary of weather. On the other hand, climate is neither a "expected weather" because this is only one aspect of the distribution. What they really want to emphasize is that the statistical distribution also includes the extreme which is really the dangerous aspect of climate rather than the average. A possible consequence of this type of reasoning is that the climate for a generic year (or period of time) cannot simply be traced to the histogram of weather from the previous years. According to some basic statistical rule, this would require the years to be interchangeable while each climatic year is unique.

Based on these premises, Rougier has invented the concept of climate simulator (Rougier and Goldstein 2014). A climate simulator following the previous definition is a numerical device that turns climatic forcing into weather but in practice a climate simulator has as main element a climate model that could go from the extremely simple (like an energy balance model, EBM) to the most complicated (like a general circulation model, GCM). We will discuss again GCMs but what they do essentially is to integrate basic equations of fluid dynamics and thermodynamics to determine the atmospheric circulation and the associated effects. The integration is carried out with numerical methods on a three-dimensional grid. Models are run in time and utilize some forces which may include solar radiation, volcanic activity, and greenhouse gas concentration. Models cannot solve explicitly all the processes involved in the complex climatic system, so they utilize "parameterization" when the results of a process can be approximated by function of some variable already solved by the model like temperature for example. Some model utilizes something like 100 parameters and their uncertainties may result in a large spread in model prediction. A climate simulator is essentially a climate model fed with initial conditions, forcings, and a class of parameters. The output of the simulator will be mainly a function of those parameters with associated uncertainties. Rougier and Goldstein may then assume

Finally, our definition of climate seems to be consistent with current practice in climate modeling, as we describe in more detail in the following sections. A climate simulator is just that, a device for generating a family of distributions of weather. Insofar as the simulator is the outcome of many judgments, its distribution is subjective. Climate modelers do not accept one of the simulator's climates as their own but make a subjective adjustment reflecting their judgment about the simulator limitations. So we find that the practice of climate modelers is inherently subjective, and that defining climate to be a subjective distribution of weather is reasonable not just from a foundational point of view but also from a naturalistic one.

Unfortunately, this approach up to now is not very popular and the climate predictions we will discuss in the book are based on the results of running a single GCM that may be with different parameterizations.

How Climate is Studied

A climate scientist and a humanist at the end of each chapter will discuss what we intended to say and what they have understood. The humanist in this case mostly tries to familiarize with some of the new concepts that the climate scientist insists should remain in the framework of the physical sciences.
H:It is interesting the distinction between meteorology and climate studies but it is not clear that they require very different approaches.
CS: Unfortunately, I think that one of the problems of the climate research is the assumption that climate could be studied with some by product of the methods used in meteorology. For example, the climate models derive directly from models used for the weather forecast. Today, this tendency is aggravated by the idea of the so-called "seamless prediction" that pretend to obtain climate predictions from a long-term integration of forecast models. The method is an invention by Tim Palmer
H: I understand that the differences are much deeper than that. According to Tim Palmer *However, probe below the surface and it can be seen that thinking about weather and climate forecasting as essential separate activities is not as scientifically meaningful as it might seem. Indeed, I would claim that trying to think about them as separate activities is actually hindering the development of climate science. Not least, weather prediction provides clear-cut metrics of model performance and it should therefore be used in the plethora of techniques needed to help develop a model which can pass the climate Turing test.* (see Palmer 2016).
CS: As a matter of fact, even if we remain in the modeling aspect, there is difference in the philosophy. The weather forecast models solve an initial value problem while the climate models solve a boundary conditions problem. This means that while the first starts from some initial observed conditions and parameters measured in some grid point, the second assigns general external conditions (like the amount of greenhouse gases) and arrive to a final state consistent with such boundary conditions. In recent times, some people is advocating the inizialization of GCMs and we will discuss this in the coming chapters.
H: As I understand, there are also different ways to verify the results obtained in the two cases. I wonder if this is related to the climate Turing test.
CS: Well the Turing test for climate is a Tim Palmer invention to give scientific prestige to climate science. In this case, the models pass the test if you cannot distinguish between predicted and observed variables at the grid scale larger than the resolution of the model. Weather models pass easily this test. The weather forecast can be verified day by day as the data become available, while the climate predictions do not have such possibility. Only recently there have been proposals to test the climate predictions on a database that cover a few decades and are known as reanalysis.
H:The scientific character of meterology is confirmed by the fact that it has somewhat originated the chaotic theories.

CS: Not just meteorology because the discovery was the merit of Ed Lorenz, while he was studying a classical problem, that is, the convection in a fluid. However, Lorenz was a meteorologist and he applied some of the new concepts to study the predictability in meteorology. He showed that a small error in the initial conditions could propagate rapidly growing at the same time to the point that after a few days the prediction was completely unreliable. As shown later by Tim Palmer, this is not only the problem that limits the validity of the prediction but is also the limited resolution of the model that makes difficult to assimilate the data at high resolution provided for example by the satellite observations.
H: One of the critics is that while the meteorology has all the characteristics of a science, the same cannot be said for the climate that uses methods that show strong similarity with applied science. This means that its results are less trusted?
CS: We will talk about that in the following chapters but a careful examination of the paleoclimatic data gives a very bad impression on the scientific validity of the climate science. It is quite surprising that the scientific community treats these data as having a non-questionable quality. A few dissenters like Carl Wunsch affirm that for example the so-called paleo circulation of the ocean is poorly known and yet the ice ages an the theories that try to explain them are one of the most popular aspects in the scientific community. However, these works are very far from the rigor required by the physical sciences. The same thing is true for the most recent data that become reliable only in the last 50 years, while most of the debates on global warming assume that these data can be extended with the same validity to 150 years in the past. For example, if we consider the effects of the volcanic eruptions in historical times, the real hard data refer only to the last two great eruptions El Chichon and Pinatubo. Despite of this the literature is flooded with works based on nothing. In 1978, Jim Hansen and coworkers published a paper on *Science* claiming a 0.5 C drop in temperature after the eruption of Agung occurred 15 years earlier with data on the aerosol extrapolated from very indirect measurements, a questionable example of bad physics. Most recent papers based on the reanalysis data show very marginal effects.
H: It is not clear to me the origin of such scarcity of quality and quantity of data, especially when one thinks that the most recent data are obtained from satellite observations or very sophisticated methods.
CS: You must consider that climatic database are a direct consequence of the data produced by the meteorological networks. These data had important development starting from the second world war. The satellite data started to come in at the beginning of the 70 s but these are mostly non-calibrated data and so they lack the quality required by the climatic data. Remember that we are talking about temperature changes of the order of 1 hundredth of degree per year and for such precision it is just out of question for satellite data. This is the reason why instruments have been proposed to produce calibrated data that can be compared with international standard especially for spaceborne observations. Such proposals have always found resistance from the

meterological community. This because the limitation in the financial resources produces a very hard competition between the organisms that manage the meteorogical data and the "scientific community". This most likely is the one that succumb as recently happened for the CLARREO project. Another example of the imprecision in the data that have hindered the proliferation of theories deals with the method of the oxygen isotopes used to reconstruct the ice sheets' volume. The precision of this method depends on the reconstruction of the depth of the cores that is a function of the sedimentation rate. However, this depends on the oceanic paleo circulation that, we have seen, is still in its infancy.

H: In the last part, it is suggested that the climate sensitivity could be estimated studying the fluctuations of the system. I understand that the climate system is made of fast component (like the atmosphere) and slow components (like the ocean) . How this method distinguishes these two components

CS: There is no way to separate the two components. In practice, the study deals with the fluctuations as are integrated by the complexity of the climatic system. Unfortunately, it is a method that has given very contradictory results.

References

Broecker, W. S. (1991). The great ocean conveyor. *Oceanography*, *4*, 79.

Broecker, W. S. (2010). *The great ocean conveyor*. Princeton University Press.

Callen, H. B., & Greene, R. F. (1952). On a theorem of irreversible thermodynamics. *Physical Review*, *86*, 702.

Canty, T., Mascioli, N. R., Smarte, M. D., & Salawitch, R. J. (2013). An empirical model of global climate-Part 1: a critical evaluation of volcanic cooling. *Atmospheric Chemistry and Physics*, *13*, 3997.

Charney, J. G. (1947). The dynamics of long waves in a baroclinic westerly current. *Journal of the Meteorological*, *4*, 135.

Dyson, G. (2012). *Turing's cathedral*. Vintage Books.

Fujiwara, M., Ibino, T., Mehta, S. K., Gray, L., Mitchell, D., & Anstey, J. (2015). Global temperature response to the major volcani eruptions in multiple reanalysis data set. *Atmospheric Chemistry and Physics*, *15*, 13507.

Gordon, A. L. (1986). Inter-ocean exchange of thermohaline water. *Journal of Geophysical Research: Oceans*, *91*, 5037.

Hanel, R. A., Conrath, B. J., Jennings, D. E., & Samuelson, R. E. (1992). *Exploration of the solar system by infrared remote sensing*. Cambridge University Press.

Hansen, J. E., Wang, W. C., & Lacis, A. A. (1978). Mount Agung eruption provides test of a global climatic perturbation. *Science*, *199*, 1065.

Hoffman, P. F., & Schrag, D. P. (2002). The snowball Earth hypothesis:testing the limits of global change. *Terra Nova*, *4*, 129.

Huang, X., Farrara, J., Leroy, S. S., Yung, Y. L., & Goody, R. (2002). TCloud variability as revealed in outgoing infrared spectra: Comparing model to observation with spectral EOF analysis. *Geophysical Research Letters*, *29*. doi:10.1029/2001GL014176.

Kirk-Davidoff, D. B. (2009). On the diagnosis of climate sensitivity using observations of fluctuations. *Atmosphere Chemical Physics*, *9*, 813–22.

Leith, C. E. (1975). Climate response and fluctuation dissipation. *Journal of the Atmospheric Sciences*, *32*, 2022.

Lorenz, E. (1963). Deterministic nonperiodic flow. *Journal of the Atmospheric Sciences*, *20*, 130.
Lovejoy, S. (2013). What is climate? *EOS*, *94*, 1–2. Permission from the author
Lowell, T. V., Fisher, T. G., Comer, G. C., Hajdas, I., Waterson, N., Glover, K., et al. (2005). Testing the Lake Agassiz meltwater trigger for the Younger Dryas. *EOS*, *40*, 365.
Murton, J. B., Bateman, M. D., Dallimore, S. R., Teller, J. T., & Yang, Z. (2010). Identification of Younger Dryas outburst flood path from Lake Agassiz to the Arctic ocean. *Nature*, *464*, 740.
Nebeker, F. (1995). *Calculating the weather, meteorology in the 20th century*. Academic Press.
Palmer, T. N., Doblas-Reyes, F. J., Weishmeier, A., & Rodwell, M. J. (2008). Toward seamless prediction calibration of climate change projections using seasonal forecasts. *Bulletin of the American Meteorological Society*, *89*, 459. © American Meteorological Society. Used with permission.
Palmer, T. N. (2016). A personal perspective on modelling the. *Proceeding of the Royal Society A*, *472*, 20150772 unrestricted use.
Pollack, J. B., Toon, O. B., Ackermann, T. P., McKay, C. P., & R. P., Turco,. (1983). Environmental effects of an impact generated dust cloud: implications for the Cretaceous-Tertiary extinction. *Science*, *219*, 287.
Rapp, D. (2010). *Ice ages and interglacials: measurements, interpretation and models*. Springer.
Rougier, J., & Goldstein, M. (2014). Climate simulators and climate projections. *Annual Review of Statistics and its Application*, *1*, 103. Reproduced with permission of Annual Review © by Annual Reviews
Seager, R., Battisti, D. S., Yin, J., Gordon, N., Naik, N., Clement, A. C., et al. (2002). Is the Gulf stream responsible for Europes mild winters? *Quaretly Journal of the Royal Meteorological Society*, *128*, 2563.
Smagorinsky, J. (1983). *The beginnings of numerical weather prediction and general circulation modeling: early recollections*, in B. Saltzman (Ed.), *Theory of Climate* (Vol. 25). Advances in Geophysics. Ac. Press
Stephenson, D. B., Collins, M., Rougier, J. C., & Chandler, R. E. (2012). Statistical problems in the probabilistic prediction of climate change. *Envirometrics*, *23*, 364.
Turco, R. P., Toon, O. B., Ackermann, T. P., Pollack, J. B., & Sagan, Carl. (1983). Nuclear winter: Global consequences of multiple nuclear explosions. *Science*, *222*, 1283.
Toon, O. B., Pollack, J. B., Ackermann, T. P., Turco, R. P., McKay, C. P., & Liu, M. S. (1982). *Evolution of an impact-generated dust cloud and is effects on the atmosphere* (p. 190). Special Paper: Geological Society of America.
Wunsch, C. (2004). Gulf stream safe if wind blows and Earth turns. *Nature*, *428*, 601. Reprinted by permission from Macmillan Publishers Ltd.
Wunsch, C. (2001). Global problems and Global onservations. In G. Siedler, J. Church, & J. Gould (Eds.), *Ocean circulation and climate*. Ac: Press. With permission from Elsevier.
Wunsch, C. (2010). Towards the understanding of the paleocean. *Quaternary Science Review*, *29*, 1960. With permission from Elsevier.

Chapter 3
Modeling the Environment

3.1 Introduction

Our objective is to show that there is basic philosophical problem about the verification and the credibility of climate predictions. The main tool used to make such predictions is the so-called GCM, an acronym that can be interpreted as global climate models or global circulation models. The first interpretation may be more correct because GCM are not just models that simulate the circulation of the ocean and atmosphere but something much more complex. We will then stick to the more general definition of models of the environment. In them, we will include also the numerical weather prediction models (NWP). This approach is also suggested by two books one by Beven (2008) of the University of Lancaster (UK) and the other by Eric Winsberg of the University of South Florida (Winsberg 2010). Beven is a hydrology professor (the science that forecasts floods) while Weisberg is a professor of philosophy and this could be our perfect combination. Actually, the point of view expressed by Beven could be more interesting because it is related to the practical applicability of models. Beven affirms something that could be very appropriate also for GCM and that is computers may produce huge amount of data while the real question is what these data represent and why we should believe in them.

We will not describe in any detail what the GCM are but rather discuss if they can be conceived as "experimental" tools that is if they can substitute in some way measurements that are used in other scientific sectors. The central point remains to establish if the data produced by such models are credible from a strictly scientific point of view that is if they can be verified. May be worth to note that this is a sector where many unleashed non-experts fill the media (but sometimes even professional journals) with comments, conjectures, and considerations. Some of these people may have never touched a computer but this is not a justification to ignore their conclusions. As a matter of fact, the most authoritative outsider is the famous science philosopher Karl Popper author of many interesting and inescapable concepts that we will discuss in this chapter.

G. Visconti, *Problems, Philosophy and Politics of Climate Science*,
Springer Climate, https://doi.org/10.1007/978-3-319-65669-4_3

How it is widely recognized, by all those that work with this subject, the climate models have many problems but at the same time are the best of what we have at the moment. This simple statement corresponds to what Beven calls **pragmatic realism**. The climate models are derived from the weather prediction models and if all the shortages contained in the present climate models will be corrected they will be able to cover all possible ranges of variability of the climate system. The pragmatic realism has been very useful to all the modeling of the environment especially for the hydrology and for the climate. Both fields require huge computational resources and are necessary for the floods warnings following heavy rain (hydrology) or the management of the energy resources (climate). The pragmatic realism is only a version of the realism theories in the philosophy of science and is the most useful version to handle the modeling practice. The important assumptions that justify the pragmatic realism need not distract us from our effort that must remain within the scientific realm. Another clarification has to be made about the reasons to embarque in a terrain that may look a little bit too philosophical. The main objection is that someone claims that models are not valid and consequently not respondent to the scientific rigor. On the opposite end, someone arrives at the incredible conclusion that models may constitute the base for crucial experiments and this may be little bit misleading.

3.2 Models Like Experimental Tools

There is philosophical current named **instrumentalism** which assumes that models are simply tools without making any hypothesis if they represent reality. This point of view is quite similar to the engineering vision of the environmental modeling. As a matter of fact in the engineering practice, models are used to make prediction about the working of structures of some machine. All the predicted quantities represent the "best estimate" of the interested quantity. This implies that when using a model, the engineer knows its limits and so the prediction he makes is the best he can do with the means he holds. Actually, the philosophical assumption at the base of the instrumentalism is that all the scientific theories of the past have revealed to be false or wrong and there is a reason to believe that this may be true also for the present ones. This does not mean that prediction is not possible but more simply that the predicted quantity is not to be confused with the "absolute" reality.

This philosophical condition may be shared because in the scientific community which deal with the climate, models are substituting the observable reality. Sometimes ago, Richard Lindzen, one of the few dissident on the global warning issue wrote (Lindzen 2012):

In brief, we have the new paradigm where simulation and programs have replaced theory and observation, where government largely determines the nature of scientific activity, and where the primary role of professional societies is the lobbying of the government for special advantage.

Lindzen ideas followed a number of papers from the operational level to the philosophical realm. The first category includes the work of Antonio Navarra, James

Kinter e Joseph Tribbia, (Navarra et al. 2010) while at the institutional level, we could include Parker (2009) and later the book edited by Duran and Arnold (2013). It is to notice that Navarra et al. never mention the "philosophical papers" and the book appearing later never mentions the climate simulation. On the other hand in the Navarra et al. paper there is some attempt to hit the philosophical arguments.

A strict application of the scientific method requires a process of isolation of constituent subsystems and experimental verification of a hypothesis. For the, this is only possible by using numerical models. Such models have become the central pillar of the quantitative scientific approach to climate science because they allow us to perform "crucial" experiments under the controlled conditions that science demands. Sometimes crucial experiments are recognized as such at the design phase, like the quest for the Higgs boson currently going on at the European Organization for Nuclear Research [Conseil Europeen pour la Recherche Nucleaire (CERN)]. Other times it is only in historical perspective that some experiments are recognized as truly "crucial". This was the case of the 1887 test by Michelson and Morley that rejected the hypothesis of the existence of the "luminiferous aether", an undetected medium through which light was deemed to propagate. Their result led to a reformulation of a physical theory of electromagnetic radiation and to special relativity and the invariance of the speed of light. "Crucial" experiments test competitive theories and the most successful one is finally selected.

The thesis of this paper is to show that climate models are a useful surrogate of the scientific experiments.

But this is only the beginning of the singular opinion contained in the paper. Classically, crucial experiments imply that from a clearly formulated hypothesis, it is possible to obtain a deduction that could be compared with the experimental reality. A few examples of crucial experiments are now part of the mythology of science: the Michelson Morley experiment that buried the idea of the ether existence or the measurement of the deflection of light by gravitational bodies performed by Eddington that constitutes one of the proofs of the general relativity. This has been a powerful procedure revealed to be essential in establishing the real nature of science. But this principle is too restrictive for the applied sciences (as the atmospheric sciences) for which one, the principal motivation remains the societal values and a precision enough for its practical scopes, rather than the most appropriate for confirming the law of nature. Actually, there is not a great distinction between applied and fundamental sciences if not for the motivation and so in principle, there is not a great limitation to the use of crucial experiments. This is a topic we will discuss extensively but the atmospheric sciences and the modeling aspect are not susceptible to any crucial experiments or in the Popperian sense to "falsification". Karl Popper had defined a procedure such that theories could only be falsified with experiments that could not confirm the theory. In the case of models however in order to simplify the representation of some process, the same processes are not described in detail but are "parameterized" that have become function of only a few variables that can be adjusted during the simulation. The one just described in itself (as someone suggested) constitutes a falsification so that it becomes a little bit embarrassing to falsify something that has been already falsified.

A possible example of crucial experiments in the atmospheric science could be the airborne campaign to get the data on the ozone hole (Christie 2000). In this case, the competing theories were the chemistry cause as compared to the dynamics effects. However also in this case, most of scientific evidence already pointed out to the first hypothesis.

The work of Navarra et al. has other ideas, some of which could be rather dangerous. On one hand, it tends to change the problem of climate prediction from a boundary value problem to an initial value problem. Up to now, weather forecast was a typical initial value problem while the climate prediction was a boundary value problem. The thesis of the above paper is that with the drastic improvement of the computer performances also the climate prediction may become an initial value problem. This will require computer performances of the order of petaflop (a thousand trillion of operations per second) or of the exaflop (a billion of operation per second). This would imply the possibility to increase the spatial resolution down to 1 km and then to assimilate, for example, convective processes at a level of a single cloud. However, a problem will remain that is central to the nature of climate science. To predict the climate in 100 years, one need to predict for example what would be the economy in the single country and the politics of those countries including revolutions, terrorist attacks or more simply the evolution of the corruption. Again one of the main motivation of the climate science is their social values which at the same time could imply large uncertainties.

Finally, there is another dangerous suggestion about what they call industrial computing

Industrial computing and numerical missions will rely on that capability even more to allow climate science to address problems that have never before been attempted. The global numerical climate community soon will have to begin a proper discussion forum to develop the organization necessary for the planning of experiments in the industrial computing age.

The thesis being that, similarly to what has been done by the astronomical communities and the particle physics communities, also the climate community could build facilities like the LHC (Large Hadron Collider) or the Hubble Telescope or the European Southern Observatory (ESO) to reach the level of Big Science.

As noted by Roger Pielke sr. in his blog (https://pielkeclimatesci.wordpress.com/2010/04/28/comments-on-numerical-modeling-as-the-new-climate-science-paradigm/)

The proposal put forth in Navarra et al. (2010), *if adopted, would concentrate climate modeling into a few well-funded institutions, as well as focus the use models for multi-decadal predictions of the real (in which we do not, of course have observational validation data), rather than as a tool to test scientific hypotheses against real world observations. Policy decisions will be made from these unvalidated model predictions (has they have already been made based on the global average and regional scale from the IPCC multi-decadal model forecasts). This is a path that will likely lead to the eventual discrediting of the climate science community who participates in this activity if, as I expect, the regional multi-decadal regional (and even global average forecasts) generally fail to show skill in the coming years.*

Besides this consideration, we must take into account the political control of the centers that make the prediction confirming the prophetic words by Richard Lindzen that we mentioned before.

There are two additional considerations that make things worse. On one side, the position of the American Meteorological Society expressed by the Editor of the *Bulletin*.

If climate science develops the way Navarra et al. suggest will this be proof that the age of numerical experimentation has matured? Perhaps so. A science shaped by Franklin and Lorenz's critical experiments is now a critical experiment itself a—test of the viability of science when it is dependent on numerical modeling for methodology. For better or worse, the result of this grand experiment the—the very state of climatology—will forever be ingrained in popular consciousness.

The other side is that models are not in a very good shape. The scientific question asked about the climate prediction is: how the experimental observation can give some constraints to the prediction? This cannot be accomplished by increasing the computer power but looking at existing data and devising new experimental procedures. Part of the scientific community is moving in such a direction.

The problem raised by Navarra et al. was not new as we already noticed. In 2005, the Presidential Information Technology Advisory Committee (2005) issued a report in which it was stated:

Together with theory and experimentation, computational science now constitutes the "third pillar" of scientific inquiry, enabling researchers to build and test models of complex phenomena.

This raised many objections and a very authorative one was by the editor in chief of the *Communication of the ACM (Association for Computing Machines* Moshe Vardi which made a very simple consideration (Vardi 2010).

Vardi observes that any application of a mathematical theory requires computation. For this reason computation has been always an integral part of science. He argues that only the scale of computation has changed. Once upon a time computation was carried out by hand while today computation requires more and more sophisticated machinery. Not only that because the nature of theories has also changed. For example Maxwell equations constitute an elegant simple model of reality. On the other hand the climate science implies the solution of very complex computational model (i.e., GCMs). The conclusion for Vardi is that while previous scientific theories were framed as mathematical models, today theories are often solved as computational models.

And again Vardi assumes that computation is rather an integral part of experimentation.

Vardi then observes that computation has been always part of the experimentation because it is needed to handle the huge amount of data produced by complex experiment. He reports the example that one of the experiments of the Large Hadron Collider generates 40 terabytes of raw data per second, a volume that cannot be stored and processed. Such large amount of data do require advanced computation and sophisticated data analysis techniques.

On the other hand, the Society of Industrial and Applied Mathematics (SIAM 2017) states that

Computation is now regarded as an equal and indispensable partner, along with theory and experiment, in the advance of scientific knowledge and engineering practice. Numerical simulation enables the study of complex systems and natural phenomena that would be too expensive or dangerous, or even impossible, to study by direct experimentation. The quest for ever higher levels of detail and realism in such simulations requires enormous computational capacity, and has provided the impetus for dramatic breakthroughs in computer algorithms and architectures. Due to these advances, computational scientists and engineers can now solve large-scale problems that were once thought intractable.

This statement is rather dangerous and contains some contradiction. One side admits that it is impossible to study with direct experimentation some complex systems like climate. This corresponds to give up in planning experiments over several years (like CLARREO or other space observations) because they are too costly but this is an attitude that is not common to other sciences like physics. Imagine to substitute the LIGO (Laser Interferometer Gravitational-Wave Observatory) detector with a simulation of the gravitational signal. Notice also the distinction between computational science and computer science. The first is about simulation while the latter is strictly about the computational machine.

There are all signs that legs are growing on science just because some industrial interests are growing.

As we mentioned at the beginning, academia had started also a debate on these questions with strongly opposing views and we think the most illuminating paper is the one by Keller (2002) that for the first time wrote the word *computational physics*.

Over the last half a century, a new domain of physical science has come into being that is widely recognized as different from the older domains of both theoretical and experimental physics, and that has accordingly warranted a new designation, namely "computational physics." Computational physics is simply a term referring to the use of computer simulation in the analysis of complex physical systems, and, as such, it is unquestionably both new and distinctive.

Tracing back to the origin of computational physics, it is clear that computers (or computational techniques) have to be regarded as new tools and as a matter of fact, a paragraph in the paper reads *Computers as Heuristic Aid: "Experiments in Theory"*. All the reported examples show that experiments as we always have intended and numerical experimentation remain two well distincts matter. They deal with all those problems that were not mathematically tractable and the "experiment in theory" in computer simulation

were 'experimental' in the same sense in which a thought experiment was 'experimental' (or in which repeated games of chance were 'experimental'), different only in that the computer permitted the working out the implications of a hypothesis so rapidly as to rival the speed of thought, and certainly vastly faster than any of the traditional means of computation that had been available. They extended the mathematician's powers of analysis and, as such, ought be as valuable for solving problems in pure mathematics as in mathematical physics. In no sense were they to be confused with actual experiments ('experiments in practice'?) on which confirmation of theory depended.

Notice the clear distinction between the actual experiment called "experiments in practice" and the "experiment in theory." Keller elaborates further by distinguishing first between theory and simulation and then between real experiments and computer experiments. Most of the time at the least in the first case, we are witnessing to a real futile debate. If you have a theory on some phenomena, you may write down equations that sometimes you cannot solve analytically so you recur to simulation. The results will depend on the assumptions you made in translating your equation in a computer program. Still you can regard all this as theory. It is very hard to accept the conclusion on the possible equivalence between computer observations and observation based on experimental measurements. Keller goes on illustrating a real novel application of computer simulation which deals with cellular automata that up to now have had only limited application to the geophysical sciences like in hydrology studies.

In another notable reference (Haile 1992), there an example is reported that shows how limited is the vision of the practioners of the trade. The example refers to a presentation of the *Voyager* planetary probe when the pictures taken of Jupiter were compared with some simulation of the atmospheric dynamics. The thesis was that simulation was less costly and more accurate than the picture and Haile easily remarks .... *The CEO could be faulted on two counts: not only did he erroneously claim that a simulation can supplant reality, but he also confused a photographic image with reality*. We like to conclude mentioning yet another more recent point of view on this debate. Barberousse et al. (2009) point out that even if simulations do not involve any measurement interactions they do generate new data about empirical system, similarly to what field experiment does.

No one of these papers ever mention the problem of climate modeling or simulation but it is quite clear that such a practice has several drawbacks. The simulation practice, especially in material sciences and synthetic biology, has obtained impressive results because in one case there is plenty of data or in the other, to follow Keller's words,

It is employed to model phenomena which lack a theoretical underpinning in any sense of the term familiar to physicists—phenomena for which no equations, either exact or approximate, exist (as, e.g., in biological development), or for which the equations that do exist simply fall short (as, e.g., in turbulence). Here, what is to be simulated is neither a well-established set equations of differential equations (as in Ulam's "experiments in theory") nor the fundamental physical constituents (or particles) of the system (as in "computer experiments"), but rather, the phenomenon itself. In contrast to conventional modeling practices, it might be described as modeling from above.

On the other hand, the has its physical laws but the interactions of the different elements are poorly known and each element is very complex. Beside the data available are scarce and of poor quality. This suggests that a possible way out is to simplify the system, that is to follow the road of reductionism. We must keep in mind that the effort to modeling climate is to give government indication on how to manage our future. The comprehension of the climate system may be a different problem. How affirmed by Haile

the validity of the model and theory would be tested by comparing prediction with experimental measurements...However the connection between input and output observables would remain implicit in the mathematical apparatus used to make the prediction

So actually, we can separate the political problem (predict the future climate) from the scientific one (comprehend how the system works).

3.3 The Validation of Models

In the discussion below, we will follow closely Beven but also Naomi Oreskes. In practice, the validation of models corresponds to make a test on the model that has been calibrated for a particular situation. The test could, for example, refer to a period different from that of the calibration. From a strictly philosophical point of view, the term validation or verification implies a value of truth that is not possessed by any model. Even worse is that Naomi Oreskes of Harvard University asserts that all the environmental models cannot be validated because they refer to a natural system. Stated in another way no model is completely confirmed by the data and no model is completely refused by the data.

The scarcity and quality of the data is well emphasized by Wunsch (2010)

From one point of view, scientific communities without adequate data have a distinct advantage: one can construct interesting and exciting stories and rationalizations with little or no risk of observational refutation. Colorful, sometimes charismatic, characters come to dominate the field, constructing their interpretations of a few intriguing, but indefinite observations that appeal to their followers, and which eventually emerge as "textbook truths".

This is because models are not unique in representing a certain situation and can only obtain a conditional confirmation. This means that the confirmation may depend on some calibration parameters or on other auxiliary conditions or that the model results may depend on the data used in the calibration. Another problem arises in relation to the so-called open system. An open system is the one that needs specific parameters or external conditions. When a model or an hypothesis are tested such external parameters must be kept under control. If the test turns out to be negative we never know if this results from a bad control of the external parameters or if the theory is incorrect. Sometimes, for example, a correct theory could not be confirmed because of the inequadecy of the instrumentation. The most famous case may the heliocentric theory of Copernicus (Oreskes 2003). The detractors of this theory at that time made the hypothesis of a "small" universe with Earth orbit of dimension similar or comparable to the stellar distances. Based on this hypothesis, they expected to measure changes in the parallax (that is the angle subtended by a star) relatively large in the course of a year. Once the measurements were made these gave negative results because the sensitivity of the instrumentation was completely inadequate.

Something like 400 years would be required to close the gap. Naturally also the other hypothesis (Earth orbit of dimension comparable to that of the universe) was incorrect.

According to Oreskes, the systems could be open based on three different criteria. First of all, we have to confront the problem to establish which are the most important processes to include in the model and which are to be discarded. Then there is the problem of how to translate them in mathematical equations and this has to do directly with the parameterization issue. Finally (this being related to the first question) which variables are to be considered.

The question to ask is the following: if the environmental models cannot be validated (or the word validation is devoid of significance) how GCMs can be used to make meaningful prediction? This is not a simple philosophical question because even taking apart the matters related to the climatic change (influenced also by rancor and academic diatribes) models are used to forecast floods or air and water quality. In particular, a relevant problem is that of the uncertainty of the forecast. One of the most cited episodes refers to the Red River flood in 1997 (Pielke 1999). A few weeks before the National Weather, Servive had issued a warning that water would reach a level between 14.47 and 14.93 m. Local authorities then strengthened the margins for a flood up to 15.5 m. But the real floods reached a level of 16.5 m causing a damage of 2 billion dollars. The danger was evident from the message and the fact that the real flood was about 1.5 m above the initial forecast and was attributed to an erroneous forecast. This episode exemplifies that it is essential to accompany the forecast with an adequate uncertainty. If we transfer this criterion to the climate prediction, we enter a very dangerous terrain. We know approximately the predicted variables let alone the errors.

3.4 The Responsibility of Scientists

The question of model uncertainties is of such importance that deserves more discussion. In any case, these uncertainties must be considered as part of the enormous responsibility of the scientists when presenting the possible consequences of the climatic changes. Many times, they emphasize the most dramatic consequences of such prediction (melting of the polar ice with the subsequent destruction of polar bears ecosystem, etc.) without mentioning that as in the case of the North Dakota floods the consequences could be even worse. As a further example, the ozone hole was never predicted by the models before it was discovered. The climate models give predictions that still have a poor resolution. For example, the so-called regional climate models give predictions with considerable uncertainties and errors. Very often we read details on the prediction of precipitation and temperature up to the second decimal digit that in a few occasions may look a little bit humoristic. In the words of Palmer (2016)

These matters become even more important if we focus on the regional climate response to increased atmospheric concentrations of greenhouse gases. A crucial aspect of such response is in climate extremes: for example, persistent circulation anomalies which can bring drought and extreme heat for some regions and seasons, and extreme flooding for other regions and seasons. This makes it clear that estimates of future climate change require us to understand the impact of atmospheric CO_2 *on the dynamics of the, and not just its thermodynamics On the other hand before the uncertainty issue there is something more important related to predictability of weather and climate*

On the other hand, before the uncertainty issue, there is something more general related to the predictability of weather and climate. The problem hides a matter of principle (may we call it philosophical?) which has been discussed by the same Karl Popper many years ago in a Compton Memorial Lecture given at the Washington University in 1965 with explicit title "Of clouds and clocks" published in 1972in the book *Objective Knowledge* (Popper 1972). In this work, Popper affirms that the clouds are non-predictable objects while clocks are representative of completely deterministic systems. Apparently, this boils down to the debate between deterministic believers and non . The former claim that the knowledge of the laws of some particular phenomena and of the initial conditions is enough to study the evolution of the system in all the subsequent instants of time. From this point of view, all the clouds are clocks. As a matter of fact, the deterministic people claim that the difference between clouds and clocks is simply that for the latter we do not know the laws that regulate their origin and growth. As it is affirmed by Henk Tennekes (another model skeptic but otherwise a first-order scientist) (Tennekes 1992).

Or, in meteorological terms, a perfect model of the atmosphere, initialized with perfect data from an observation network of infinite resolution, and run on an infinitely powerful computer, should in principle produce a perfect forecast with an unlimited range of validity

It is quite singular that the work by Edward Lorenz that originated the chaos theory is from the year 1963 and it was not sure whether Popper has read it before the lecture although some hints may suggest a positive answer. This is contained in the book *The Open Universe* (Popper 1988).

The fundamental idea underlying 'scientific' determinism is that the structure of the world is such that every future event can be rationally calculated in advance, if only we knows the laws of nature, and the present state of the world. But if every event is to be predictable, it must be predictable withanydesireddegreeofprecision: for even the most minute difference in measurement may be claimed to distinguish between different events.

In some way, Popper seems to refer to the Lorenz discovery that some differences in the initial conditions may result in predicted trajectories which are completely different. Not only that because Popper introduces another concept that may result to be more stringent for the climatic prediction that he calls principle of accountability.

Scientific determinism requires the ability to predict every event with any desired degree of precision, provided we are given sufficiently, precise initial conditions. But what does "sufficiently" mean here? Clearly we have to explain "sufficiently" in such a way that we deprive ourselves of the right to plead—every time we fail in our prediction—that we were given initial conditions which were not sufficiently precise.

In other words, our theory will have to account for the imprecision of the prediction: given the degree of precision which we require of the prediction, the theory will have to enable us to calculate the degree of precision in the initial conditions that would suffice to give us a prediction of the required degree of precision. I call this demand the principle of accountability

In practice, Popper claims that it is useless to blame a failure in the prediction to the scarcity of computational resources or to the errors related to approximate measurements. The scientist must estimate in advance the precision of its prediction based on the data and the means he has in hand. Also Popper says

The method of science depends on our attempts to describe the world with simple theories. Theories that are complex may become untestable, even if they happen to be true. Science may be described as the art of systematic over— simplification: the art of discerning what we may with advantage omit

The reference to models are even too clear but the enunciated principles are hardly transferable to the practical cases. As a matter of fact even in meteorology what it is called forecast skill is computed with respect to observations or to a reference model. We should be then prepared to make benchmark measurements or build a reference model.

There is another important point that needs to be mentioned about uncertainties and communicating uncertainties that we could call following (Parker 2014), values and uncertainties in climate prediction. The problem how social values can influence scientific research has been raised in the middle of last century by Richard Rudner and Rich and Jeffreys and more recently by Douglas (2009) and specifically by Winsberg (2012) for the climate modeling. Winsberg argues that some choices of the climate modeler are necessarily influenced by the social context like choosing which climate variable needs to be predicted. The real problem, however, is more with the evaluation of the uncertainties in the prediction. Some people even assign probabilities to a particular prediction like the raising of temperature in a certain time interval. Sometimes, they even recommend an acceptable range of probability. In particular (Parker 2014) observe

Estimates of uncertainty are themselves always somewhat uncertain; any decision to offer a particular estimate of uncertainty implies a judgment that this second-order uncertainty is insignificant/unimportant; but such a judgment is a value judgment, as it is concerned with (among other things) how bad the consequences of error (inaccuracy) would be; hence even decisions to offer coarser uncertainty estimates at least implicitly reflect value judgments (see, e.g., Douglas, 2009, p. 85). A variation on this reply argues not just that uncertainty estimates always implicitly reflect value judgments but also that scientists ought to explicitly consider how bad the consequences of offering an inaccurate depiction of uncertainty would be and are remiss if they fail to do so (see ibid, Chap. 4). Douglas (ibid.), for instance, argues for the latter by appeal to agents' general moral responsibility to consider the consequences of their actions.

We are back to the problem of the flood prediction and to the responsibility of the climate modelers. We have the sensation that most of the reports issued on the future possible climate change look very much aseptic while many times imply devastating consequences. We think that not only probabilities should be assigned to the prediction but also a very accurate discussion on the implications of changes of some thermodynamic quantity, like temperature (Palmer 2016). In most cases, such consequences are known only vaguely especially on the generation of extreme events. This is the real dilemma that faces the climate science and it is just for this lack of knowledge that social values somehow fill the scientific gap.

3.5 The Falsification of Models

Karl Popper accurately avoided the problem of validation or verification of a theory saying that a theory must be judged based on its capacity to represent the reality. Nothing on the models but a few years ago David Randall of the University of Colorado and Bruce Wielicki from NASA Langley Research Center (Randall and Wielicki 1997) have resumed this problem in connection with another Popper's idea, the falsification. Popper sustained that a theory or a model (even if it was never mentioned as such) can be scientifically accepted if it is "falsifiable" that means that once compared with the experimental data the theory can be invalidated. This concept implies that no theory could be assumed to be correct because it cannot be excluded that in the future new data or measurements could disprove it. First of all, Randall and Wielicki clarified that a model includes usually a theory and this is true also for the numerical models. Models and theories allow to make prediction on the results of a measurement. However, when a modeler talks about data rarely he thinks in terms of falsification. On the contrary, they have some typical jargon. For a modeler, validation means that the data are needed to show that a model works. As we have mentioned before a model can be falsified but cannot be demonstrated that it is right. Another magic word is "tuning". For a modeler, tuning means to adjust some parameter of the model to improve the agreement between the prediction and the measurement. In practice, the modeler needs the data to tune his model. Finally, modelers also use the term calibration and that is the data are used to calibrate the model. As a matter of fact the meaning of calibration is similar to that of tuning except for the fact that in the first case it is used with positive connotation and the second with a negative one.

Often models are under the Damocles sword of the empiricism and frequently this is a very bad empiricism. Again very often we heard about "empirical" solutions based mostly on personal intuitions and not on the data. On the other hand, the empirical solutions should be universal and not established case by case. Finally, all the empirical solutions and the relative parameters should be implemented in the model before the prediction runs. What actually happens is to adjust the parameters after the results of the run. This is the most detrimental example of tuning because it is a trick that masks possible wrong results by the model. On the other hand, this is very contradictory because the most important scientific results are to show that the model is wrong so the tuning prevents the possibility to understand scientifically the problem.

The procedure of applying the falsification to the models has very important implications. If the model cannot be falsified coherently, it had to be abandoned but many think that this criteria could be unbearable by the community. A similar criteria could be even worse for complex models. Such models depend on many parameters and many parameterizations and in case of model failure, it is very hard to pinpoint the cause of the failure. On the other hand, even a model that compares well with the data could mean that some bad parameterization is compensated by other parameterization.

An important contribution by Eric Winsberg is when he observes that models are objects "analytically impervious" that is it is impossible for his creator (or creators) to understand the reason why a particular prediction is right or wrong. The other important aspect is that of the entrenchment that is the model is a victim of its own history. Models are developed and adapted to specific circumstances, to a number of constraints and their histories leave indelible traces on them. It seems to emerge that philosophy is a mean to emphasize that the main road is the scientific method while the different tricks used by the models take us more and more away from that path. Another important conclusion is that before deciding if the climate science can be counted as "science" the comparison with data remain essential. In the case of climate studies, this point is not obvious because of the quality of the data and because the methods used in the comparison are not objective.

We can draw a definitive conclusion and that is that climate predictions should use more observed data. If so you need to control the discussion so that it clearly moves toward this unique conclusion. But that is not enough. Even with more observations, one has to have some understanding of which observations, how much they can improve the prediction, and how you know that it has been improved, particularly when there is such a problem with "verification". On the other hand, the assumption that a forecast is attempting to define a state of "reality" must be abandoned. Probably very few climate modelers think this. Rather we believe, they are attempting to produce a "useful" result.

Modelling the Environment

A climate scientist and a humanist at the end of each chapter will discuss what we intended to say and what they have understood. The humanist in this case is more familiar with the many philosophical implications.

H: This chapter looks more familiar to me than to the scientific culture even if the philosophers mentioned are accused to be incompetent.

C: In this case incompetence should not be considered as an offense and in any case a different point of view must be taken into account. On the other hand we will see later that important contributions to the philosophical discussion come from scientists like Harold Jeffreys, an ante litteram astrogeophycist, or Hugh Gauch, a plant genetist. Other contribution like those of Naomi Oreskes, Eric Winsberg or Wendy Parker remain fundamentals for the discussion about model validation.

H: I have the impression that the discussion remain academic. What is the importance of such expression like *pragmatic realism* or instrumentalism while the real job is that to improve the predictions.
C: Critics to models and climate predictions came from everywhere and one of them claim that the simulation of nature does not represents reality. There is joke popular in the modeler world which says that these people believe that the real nature is the one coming out from their models. So it is very important to remember the motivations at the base of the simulations. The honest modeler knows that his model does not represent the world but at the same time is the better approximation to reality. He also knows that in 10 or 20 years new theories will substitute the present ones so that his model instruments should not be used cynically to advance in his career but are the best tools he has available at present time. Again I would like to mention Carl Wunsch
That models are incomplete representations of reality is their great power. But they should never be mistaken for the real world. At every time-step, a model integration generates erroneous results, with those errors arising from a whole suite of approximations and omissions from uncertain or erroneous: initial conditions, boundary values, lack of resolution, missing physics, numerical representation of continuous differential operators, and ordinary coding errors. It is extremely rare to read any discussion at all of the error growth in models (which is inevitable). Most errors are bounded
H: However this immense trust in the simulation has brought as indicated by Richard Lindzen to a complete inversion of priorities so that theory and observations have been substituted by simulation and software programs.
C: This is the real sore point because some people has hypothesized a crucial experiments based on models. As we well know in physics crucial experimentss are those that decide once and for all which theory is correct. The Michelson Morley experiment is the best example. On the other hand a more modern vision is that such experiments are not essential and one of the proof is the theory of relativity that would have been formulated even in the absence of the Michelson and Morley experiment.
In the geosciences the measurements in the Antarctic stratosphere by Jim Anderson could be mentioned as a crucial experiments. The results showed that the correct origin of the ozone hole were the chlorine compounds rather than the atmospheric dynamics. In the case of the climate science the same Anderson's group has proposed what could be a crucial experiments and we will talk about this later.
H: I would like to remind some data taken from the book by Paul Humphreys (a philosopher), (Humphreys 2004). He reports that the mapping of human genome required 20,000 CPU hours on the world's second largest networked supercomputer. The database contained eighty terabytes of DNA analysis. However nobody claimed the computer effort was a separate endevour from the scientific problem.
C: I agree that computer and their uses have to be regarded just as other instruments for scientific studies and are not necessarily additional scientific legs.

H: Again the problem of ethical and social values as opposed to epistemic values seems to be agitated only by philosophers while is discussed superficially by the scientist. It seems to me that it is quite obvious that all the research on predicting future climate has its basis on social values. Otherwise scientists should direct their interest rather to the basic science of climate of which climate prediction it is just one branch

C: You are quite right, In a sense this is more applied science that basic science. On the other hand is rather illusory to think that you can solve such a complex problem like climate change without improving the comprehension of the basic mechanism regulating climate, We will have occasion to talk about this later.

H: The philosophical and humanistic aspect is then important especially for the validation of models. I like to remember the contribution from Naomi Oreskes. The original ideas by Oreskes were expressed in a paper published in 1994 (Oreskes et al. 1994) but refer to a concept that was around before that. She and her co workers claimed that the more is complex the theory the more difficult is its verification. In the case of environmental model you talk about open systems those where the adjustable paremeters are quite a few.

C: The Oreskes examples on the heliocentric theories are very good. Another example very much popular in the Earth Sciences has to do with the age of the Earth as determined by the end of the ninetheenth century by Lord Kelvin. Using the data of the temperature gradient at the Earth interior and based on the hypothesis that the cooling was due to the heat diffusion, Kelvin determined an age between 20 and 100 million years. This was not right neither to the creationist (too old) nor to the Darwinist (too young for the evolution) The reason was that Kelvin had neglected the important radioactive source of heat that was quantified only in the 30's.

H: But if the models cannot be validated how can we trust their prediction? And what could the reliability of the data.

C: Climate models derive from the models used in for the weather forecast. Once the possible errors are corrected their results should be reliable. The prediction from complex models go in the same directions of those obtained from simpler models and some of these model if properly used may reproduce some of the climate of the past. The problem of data reliability is much more complex and take us back to the philosophical problem. People today is no longer available to accept coarse errors especially from science.

H: It is quite interesting that the last part of the chapter deals with a mainstream philosopher like Karl Popper. It looks that he talks about everything from the impossibility to falsify the models to the responsibility of the scientists which are ate the same time too deterministic.

C: Popper argues that valid theories must be falsifiable that is disproved by the experiments. Models are not theories but are based on them and very often to make things round few parameters in the models are tuned. This procedure corresponds to a falsification of the model so that it hard to falsify a product that has been already falsified. As for the determinism, models are deterministic instruments but they refer to a system that is chaotic in principle and so depends critically on the initial conditions. In this sense Popper had a great intuition when argues that a theory must be verified with precision as large as possible because even small deviations between experiment and theory could implies that we are talking of a different theory. On the other hand don't forget that we are talking about chaos that is a system described by a set of equations. For such system the same initial conditions give the same results. The different apparent solutions depends only on the different initial conditions. Today even in the weather forecast this is not the main problem but rather is to have a resolution large enough to include small scale phenomena.

References

Barberousse, A. (2009). Computer simulation as experiments. *Synthese, 169*, 557.

Beven, K. (2008). *Environmental modelling: An uncertain future?: An introduction to techniques for uncertainty estimation in environmental prediction*. CRC Press

Christie, M. (2000). *The ozone layer: A philosophy of science perspective*. Cambridge: Cambridge University Press.

Douglas, H. E. (2009). *Science, policy, and the value-free ideal*. Pennsylvania: University of Pittsburg Press.

Duran, J. M., & Arnold, E. (Eds.). (2013). *Computer simulations and the changing face of scientific experimentation*. Cambridge Scholars Publishing.

Haile, J. M. (1992). *Molecular dynamics simulation*. New Jersey: Wiley. Reprinted by permission from Wiley Copyright and Permissions.

Humphreys, P. (2004). *Extending ourselves. Computational science, empiricism, and scientific method*. New York: Oxford University Press.

Keller, E. F. (2002). Models, simulation, and 'computer experiments'. In H. Radder (Ed.) *The philosophy of scientific experimentation*. Pittsburgh University Press, Excerpts from Models. Simulation, and computer experiments by Evelyn Fox Keller from The philosophy of scientific experimentation, edited by Hans Radder © 2003. Reprinted by permission of the University of Pittsburgh Press.

Lindzen, R. S. (2012). Climate science: Is it currently designed to answer questions? *Euresis Journal, 2*, 161–93. Permission from the author.

Navarra, A., Kinter, J. L., & Tribbia, J. (2010). Crucial experimentss in climate science. *Bulletin of the American Meteorological Society, 91*, 343. © American Meteorological Society. Used with permission.

Oreskes, N., Shrader-Frechette, K., & Belitz, K. (1994). Verification, validation, and confirmation of numerical models in the earth sciences. *Science, 263*, 641.

Oreskes, N. (2003). The role of quantitative models in science. In C. D. Canham, J. J. Cole & W. K. Lauenroth (Eds.), *Models in ecosystem science*. New Jersey: Princeton University Press.

Palmer, T. N. (2016). A personal perspective on modelling the climate system. *Proceeding of the Royal Society A, 472*, 20150772.

Parker, W. S. (2014). Values and uncertainties in climate prediction, revisited. *Studies in History and Philosophy of Science*, *46*, 24. with permission from Elsevier.
Parker, W. S. (2010). An instrument for what? Digital computers, simulation and scientific practice. *Spontaneous Generation*, *4*, 39.
Parker, W. S. (2009). Does matter really matter? Computer simulations, experiments and materiality. *Synthese*, *169*, 483.
Pielke, R. A. (1999). Who decides? forecast and responsibilities in the 1997 red river flood. *Applied Behavioral Science Review*, *7*, 83.
Popper, K. (1972). *Objective Knowledge: an evolutionary approach*. Oxford: Oxford University Press.
Popper, K. (1988). *The open universe: An argument for indeterminism*. Rutledge. Republished with permission from Penguin. Permission conveyed through Copyright Clearance Center, Inc.
President's Information Technology Advisory Committee. (2005) *Computational science: ensuring America's competitiveness*. Washington D.C.
Randall, D. A., & Wielicki, B. A. (1997). Measurements, models, and hypotheses in the atmospheric sciences. *Bulletin of the American Meteorological Society*, *78*, 399.
SIAM. (2017). http://www.siam.org/students/resources/report.php.
Tennekes, H. (1992). Karl Popper and the accountability of numerical weather forecasting. *Weather*, *47*, 343. Reprinted by permission from Wiley Copyright and Permissions.
Vardi, M. Y. (2010). Science has only two legs. *Communication of ACM*, *53*, 5.
Winsberg, E. (2012). Values and uncertainties in the prediction of global climate models. *Kennedy Institute of Ethics Journal*, *22*, 111.
Winsberg, E. (2010). *Science in the age of computer simulation*. University of Chicago Press.
Wunsch, C. (2010). Towards understanding the paleocean. *Quaternary Science Reviews*, *29*, 1960.

Chapter 4
What Is Climate Science

4.1 Introduction

At this point it may be worthwhile to ask what is the position that climate sciences occupy in the more general field of sciences. It is not an academic question because when talking about models falsification the criteria could run from rigorous and analytic (like those of physical sciences) to more relaxed ones. Since the early days of numerical weather predictions (but even before) scientists have assumed the climate sciences to be within the realm of physical sciences. It is then reasonable to ask whether the physical sciences may contribute to increase the credibility of the climate predictions. This is not the only choice. Many years ago Richard Lewontin wrote an interesting paper about the preference of science philosophers for physics rather than (for example) biology. He explain (Lewontin 1990)

At the heart of the philosophers' preference for physics over biology is the question of uncertainty. Science is supposed to be a study of what is true everywhere and for all times. The phenomena of science are taken to be reliably repeatable rather than historically contingent. After all, if something happens only on occasional Tuesdays and Thursdays, popping up when one least expects it is like a letter from the IRS, it is not Science but History. So, philosophers of science have been fascinated with the fact that elephants and mice would fall at the same rate if dropped from the Tower of Pisa, but not much interested in how elephants and mice got to be such different sizes in the first place.

Physics and mathematics also help in the falsification process as he compares two statement

In terms of the formal calculus of propositions, the statements of science are supposed to be so-called "universally quantified" statements of the form

For all x, if x is A then x is B

rather than historical statements, which are only existentially quantified

There exists an x such that x is B.

The point, Popper tells us, is that the first kind of statement can always be falsified, by finding a single example that does not obey the rule, while we can never disprove

G. Visconti, *Problems, Philosophy and Politics of Climate Science*,
Springer Climate, https://doi.org/10.1007/978-3-319-65669-4_4

the second kind because we may have accidentally missed the cases that agree with it. So the first kind of statement is what characterizes a science, while the second kind is just storytelling.

Of course, we all know that biology is as respectful science as physics and on the other hand, there are even people that claim that science should not be regarded as a sacred cow. This problem has been raised in the 90s by the followers of a philosophical movement known as sociology of scientific knowledge (SSK) or by the so-called Edinburgh School. The main argument of SSK is that science is a social construct that is, the success of a particular theory with respect to another depends mainly on the socio economic environment where those theories are elaborated. There is a famous book by Andrew Pickering, (1984) when he was at the University of Illinois, Urbana, which is the summa of such ideas. The thesis of this book is that in physics, experiments and theories come already packed and the tradition of the experiments is such to produce the kind of data that are needed for other theories while the tradition of theories is such to generate new problems for further developments. Pickering mentions many proofs to justify his claims while the mainstream community reacted with some vigor to his observations. Strangely enough, the main confutation to the SSK position is based on the technology, if the machines work it is because their design is based on theories which are basically correct. This once again shows how the culture of some scientist is lacking because many times technology came before science (think about the Galileo telescopes). Pickering can easily rebutt such objections saying that "A machine works because it works, not because of what anybody thinks about it". Hugh Gauch in his book *Scientific Method in Practice* (Gauch 2002) referring to this discussion remarks that

In short, science's detractors are asking science's defenders for a philosophical account of its rationality, not a scientific list of its successes. This is a noble request. The bottom line is that after a philosophical justification of science has been given, listing science's successes can seal the conviction of truth; but before a philosophical defense is given and especially when a philosophical distress is already evident, various successes can be acknowledged without prompting any implication of truth.

These episodes of the so-called science wars (social sciences vs science sciences) are mentioned simply to deflate the rhetoric you often read on many books on the scientific methods. The climate science seems to be right in the middle of this debate being based on incomplete theories and experimental data of poor quality. The climate science s seem to go in the direction indicated by the Harvard's gurus, Richard Lewontin and Hilary Putnam when affirming that "science is made by scientists". (see below).

In particular Lewontin maintains that there is the necessity for the general public to understand science so that decisions are taken in a democratic fashion. On the other hand, the public, in general, does not have the necessary scientific culture. This means that very few reads directly the scientific literature with the result that most of the time the scientific information for the general public are mediated. The scientific elite, in any case, maintains an almost complete independence from the public opinion. As a consequence, their funding requests to the political establishment consider only their own interests. There is another point to be considered about basic education.

It happens often, especially considering the growing abstractness of the scientific concepts (think about the string theory), to recur to metaphors (as Lacan says a word for a word) to reach the great public but especially to print some concept in school textbooks. This means that mainly accepted ideas are transmitted rather than reserve some space to still controversial problems. Lewontin invites then to distrust what "scientists" say and to refuse the holistic approach that neglects the systematic explanation of the different components of the word. He claims that there are not "laws" in all the sectors of knowledge and that physics does not represent the paradigm for all the sciences. Again Lewontin et al. (1985).

Nowadays, even science must consider some economic constraints so that with the aim to convince people of the merits of their research, "scientists" become rhetoric about the implications of their results (typically the "war" against cancer). Rhetoric and scientific knowledge are directly connected because scientists form a cooperative enterprise which transmit through the popular press (or in general the media) what interests them more to maintain open the channels who are foraging their research. Among other things, this process cancels one of the most stimulating aspects of the scientific research because gives to the public only certainties and in any case only those who interest the scientific community. This situation can be changed only through actions that start from the lowest school level where teachers should refuse to use textbooks that oversimplify the scientific realities because often scientists exaggerate and dramatize their results in order to increase their chances at funding. Distrust always sentences like "with time and adequate funding we will understand such and such the problem."

What lesson for the climate sciences.! First of all, there are no universal law for all the sciences and physics is not a paradigm. Besides, what more clear connection exist between climatic research, funding and general public. Here, we have some kind of ecumenic organisms like the IPCC. Lewontin does not say anything about the method if not very indirectly so we need to explore a little further.

4.2 Another Bit of Philosophy

Before entering in the jungle of the hard science let us see if we can silence (or interact) with the philosophers. This is a category that does not have a specific competence so that for the high priests of science and their choir boys it remains always open to criticism. However, we want to refer here to some opinions that support what we said in the introduction and give us a hand to put the models in the right perspective. On the other hand, some of these people are active in the field of the environmental change so that his opinions may be not negligible.

In 1998 in the journal *Climatic Change*, a paper was published by Simon Shackley and his coworkers (Shackley et al. 1998) of the Center for Environmental Changes of the University of Lancaster. The title of the paper is rather explicative *Uncertainty, Complexity and Concept of Good Science in Climate Change Modeling: are GCMs the best tools*? The idea of the paper is to explain the unchallenged dominance of the GCMs in the study of climate. The answer is that the development of the science on the climate change (as the use of GCMs) and the development of concepts of global management (exemplified by the various international conferences) are contemporary and support each other. In particular, the paper explains the preference of the politicians to use very complex models that are developed only in a few centers (always remember Richard Lindzen) that need huge resources in terms of computer hardware and personnel. On the contrary, in order to have estimates of the average global effects on some climatic variables (like temperature, precipitation etc.) simpler models could be used that could be run even by the secretaries of the governative departments. There are different reasons for that. First of all, the operators of the GCMs want to maintain their privileged position for the funding which have a better chance to continually flow if the model, for example, can predict some regional effect. The same models are presented as the only tools for this kind of prediction. Complexity is then fundamental because means to involve other "sciences" that are needed to predict the greenhouse gases level. Complexity (and then impenetrability) is a kind of insurmountable barrier on one end between the politicians and the high priests (to follow Harold Schiff, author of *The Ozone War*) and on the other hand between the layman which includes both the environmentalist and the entrepreneur.

Paradoxically complexity is a very efficient defense against the easy criticism because in some sense impenetrability is apparently a warranty of an extremely sophisticated science. The development of the GCMs involves many other sciences like computer science, geology, chemistry, and so on and consequently generates a large turnover especially at the academic level. Even if it may seem inappropriate the "industry" of the GCMs has direct links with the impact assessment and then with the politics of mitigation (the reduction of the greenhouse gases) and adaptation (that nobody wants) and with an academic industry in rapid growth which is the geoengineering. The latter envisage the artificial change of the terrestrial environment to compensate for the changes produced by the human activities. Another academic sector influenced by the GCMs is the economy with the planning of industrial activities connected with the energy management and the environmental impact. In practice, the results of the research centers that manage the GCMs are capable to influence more or less directly any aspect of our future.

In another paper on the same journal Schackley himself (Shackley et al. 1999) together with others, including a famous atmospheric scientist like Peter Stone, interviewed a number of scientists on the important problem of tuning or what is known as flux adjustment. Historically, when they tried to couple the ocean models with the atmospheric models to create the sophisticated AOGCM (Atmospheric Ocean General Circulation Model), the researcher discovered that using the value of the exchange fluxes between ocean and atmosphere the AOGCM became unstable. The fluxes should then be adjusted to maintain the stability. The most surprising result of this survey was that the majority of the interviewed that were familiar with the tuning, had the idea that this practice should remain an internal question of their circle otherwise, other people could have used it as a heavy criticism. That is we have reached almost the secret society or in any case to the logic of a clan.

All these tales seem to distract us from our main direction and that is if something exist like the physics of climate. These are elements that enforce the idea that would be very hard to dismantle the perfect mechanism of the interaction between the power of the politics and climate sciences. It is very hard to convince people that there must be different ways to study the climate besides GCMs. However, one of the pivot of the physical sciences is the experimental proof as it is for many other inexact sciences. Despite everything, we need to find a way to reconduct the climate sciences within the mainstream of the experimental sciences.

4.3 Physics and Philosophy of Climate Predictions

We promised to deal less and less with philosophy but in this section we are at risk to continue. We cannot neglect a book quite well known at the scientific level (and unfortunately not known to the great public) of a plant physiologist, Hugh Gauch of Cornell University, with a title that maintains its promises, *Scientific Method in Practice*. The argumentations in the book take the part of the experimenter and the principles on which the analysis is based are the inductive and deductive logic, the probability, the statistics, the parsimony, and the efficiency. It is very instructive that the chapters on the deductive and inductive logic are separated by a chapter on the probability that can be considered a branch of the deductive logic. Not only that because the main difference between deduction and induction is what interest us more because it changes completely the nature of the problem. As a matter of fact, a model can be regarded as a deductive tool that is from a model, we can expect some kind of data. On the contrary and inductive process is such that from the observed data, we can suggest a model. This difference is implicit in that between probability and statistics.

Gauch makes a very intuitive example by defining first of all a fair coin and that is a coin which gives the same probability (0.5) for both "head" and "tail". On the other hand, a not fair coin may have for example a probability 0.6 for head and 0.4 for tail. The difference between deduction and induction in the framework of probability (deductive) and induction (statistics) can be understood by comparing the following two problems.

Problem 1. Given that a coin is a fair coin. What is the probability that 100 tosses of the coin produce 45 heads and 55 tails?

Problem 2. Given that 100 tosses of a coin produce 45 heads and 55 tails. What is the probability that the coin is a fair coin?

These two problems may look very similar but actually are completely different because they imply different directions in reasoning. The first is a deductive problem while the second is an inductive problem. The first is a problem of probability while the second is a statistical problem. This is because the difference between deduction and induction is based on principles (or better premises). First of all, the conclusion of a deductive argument is already contained in the premises (fair coin) while the conclusion of an inductive case goes beyond the premises (from the observed data of the tosses we obtain indication on the nature of the coin). The second difference comes from the fact that in a deductive reasoning given the premises defined as truth, the conclusion is derived with certainty (in the case of the coin the answer is assured).

In an inductive reasoning, given the premises defined as truth, the conclusions are verified with a high degree of probability (the probability that a coin is a fair coin). Finally in the deductive reasoning from a model, we derive the data while in an inductive process, the opposite happens because from the data, we arrive to a model. In the specific case when we know that the coin is fair, it is possible to evaluate the probability to obtain 45 heads out of 100 tosses. The inductive problem is not as simple because we must adopt a procedure that is based on the sequence of the results and gives a precise indication that the coin is not a fair one. We will talk about this in the next chapter.

Gauch again asks how induction is so diffuse in science. This is because in most cases science has to do with quantities that we do not observe directly. A trivial example is that of a melting temperature of a metal that is measured only in one place and then it is assumed to be the same for all the metals of the same type. Without induction, science would be condemned to extinction. And the induction is the key for our climate problem because in this case, we make predictions which cannot be verified by data so that the reliability of the prediction must be based on existent solid data. For this reason, we need a more sophisticated statistics like that of Bayes that we will expose next.

Just to complete this brief introduction, we should talk about the parsimony principle known also as Ockham criteria or Ockhan razor. The principle is important for two reasons. The first is that the scientific research has never produced a single conclusion without invoking the parsimony principle, that is the simplest explanation. The second reason directly concerns the climate sciences because the models, let call them economical, may help in comprehend the problem, improve the precision and increase the efficiency. Simpler model requires less data than complex models to reach the same precision. We ask whether Earth System Models of Intermediate Complexity (EMIC) may be more appropriate than complex model used for example to study regional climate. This is not the opinion of Carl Wunsch and he may be right.

Once examined these basic concepts, the problem remains that the hypothesis must be confirmed by the data and Gauch in the final paragraph of it book writes

In this figure's overview of scientific method, the key feature is that the hypotheses are confronted by data, leading to convergence on the truth about physical reality. Although the conjectured state of nature—the hypothesis or theory—may be inexact or even false, "the data themselves are generated by the true state of nature". Thereby nature constraints theory. That is how science finds truth, That is why science works.

The rhetorical style betrays the true believer in the purity of science and it is interesting to compare it with the opinion of Richard Lewontin about the scientific procedures (Lewontin et al. 1985).

We are arguing that there are two distinct questions to be asked of any description or explanation offered of the events, phenomena, and processes that occur in the world around us.

The first is about the internal logic and asks: Is the description accurate and the explanation true? That is do they correspond to the reality of the phenomena, events and processes in the real world? It is this type of question about the internal logic of science that most Western philosophers of science believe, or claim to believe, science is all about. The model of scientific advance that most scientists are taught, and which is largely based on the philosophers like Karl Popper and his acolytes, sees science as progressing in this abstract way, by a continous sequence of theory-making and testing, conjectures and refutations. In the more up to date Kuhnian version of the model, these conjectures and refutations of "normal" science are occasionally convulsed by periods of revolutionary science in which the entire framework ("paradigm"), within which the conjectures and refutations are framed is shaken, like a kaleidoscope which relocates the same pieces of data into quite new patterns, even though the whole process of theory-making is believed to occur autonomously without reference to a social framework in which science is done.

But the second question to be asked of descriptions or explanations is about the social matrix in which science is embedded and it is a question of equal importance. The insight into the theories of scientific growth hinted at in the nineteenth century by Marx and Engels, developed by a generation of Marxists scholars in the 1930s and now reflected, refracted and plagiarized by a host of sociologist, is that scientific growth does not proceeds in a vacuum. The question asked by scientist, the types of explanation accepted as appropriate, the paradigm framed, and the criteria for weighing evidence are all historical relative. They do not proceed from some abstract contemplation of the natural world as if scientists were programmable computers who neither made love, ate, defecated, had enemies, nor expressed political views.

It is from this perspective that one can see that the internalist positivist tradition of the authonomy of scientific knowledge is itself part of the general objectification of social relations that accompanied the transition from feudal to modern capitalist societies. That objectification results in a person's status and role in society being determined by his or her relation to objects, while the way in which individuals confront each other is seen as the accidental product of these relations. In particular, scientists are seen as individuals confronting an external and objective nature, wrestling with nature to extract its secrets, rather than as people with particular relations to each other, to the state, to their patrons and to the owners of wealth and production. Thus scientists are defined as those who do science, rather than science being defined as what scientist do.

What enlighting and apt reflections for the climate studies! Think about a system which includes a political organism like the IPCC, think about the relations between the fund givers, the oil companies, etc. All right Gauch may be right in his mainstream, technical rigor but the reality may look different.

Talking about induction, we cannot neglect the role of another "true scientist" like Sir Harold Jeffreys. All those who have studied geophysics know the sacred textbook *The earth: Its origin, history and physical constitution*, printed in 1924 (but then updated in various editions). This book reflects an epoch in geophysics when almost everything was made using paper and pencil because computer was not even imagined. Among his less known contribution is the idea that the general circulation of the atmosphere is also determined by what today are known as eddies, that is the large-scale turbulence. Jeffreys, however, was a great thinker on mathematical methods and in general on the scientific method. The book where he expresses such ideas is less known but it is fundamental for us and is *Scientific Inference*, printed for the first time in 1931 up to a recent edition dated 2010 (Jeffreys 2010). In this book, the statistical methods of Laplace and Bayes are reevaluated and the beginning indicates clearly the objectives that even today can be easily shared.

The fundamental problem of this work is the question of the nature of inference from empirical data so as to predict experiences that may occur in the future.

Jeffreys emphasizes the procedures that bring to the prediction rather than the comparison with the data. Gauch and Jeffleys, from this point of view, are not in contradiction while the latter is better for our purposes. In the first chapter, Jeffreys mentions a hypothetical debate between a botanist and a logic (which has inspired our discussions at the end of the chapters). In this debate, each one blames the other to get too far from the data. The implication is that the more the inference gets far from the data the more that inference loses significance. Unfortunately, climate projections also are drifting away from the data.

Jeffreys never tries to hide that his preference is for the inductive method even if sometimes this could be contradictory. As we said before, the inductive method starts from a single fact (the measure of the melting temperature of a piece of iron) to generalize it to all the iron found on the planet. However, this absolute statement cannot be proved without examine all the observations which have produced it. We have mentioned why the inductive method is such a powerful method or as the epistemologist Charlie Broad says induction is the glory of science and the scandal of philosophy. This is simply to emphasize that philosophy has precision standard very much superior to those of science. The scientists like to be right but not always necessarily and as Gauch affirms science has a safety net. Researchers work (or should work) in such a way that the nature should constraint their conclusions. Besides, scientists aim to find out the real state of the nature without never reaching it. Philosophy, on the other hand, deals with absolute concepts.

All these consideration take us to adapt the classical principles to the climate predictions. A climatic prediction must be the nearest possible to all the relevant and available data (evidence). The nearest to the data, the more the prediction will be reliable. But if we accept this rule what kind of data we should use if the prediction is for the future. How then a climate prediction can take into account the data?

4.4 Are Climate Sciences Part of the Physical Sciences?

We started out this chapter with the idea to ascertain the existence of a climate science which satisfies all the criteria of an observational science. This includes earth sciences as well as astronomy that is sciences where experimental data are mainly gathered by passive observations. Based on the experimental proofs obtained from those observations theories are accepted or rejected. The sanctification of climate science has obvious political implications because it certifies that their practioners have to be trusted by the governments when they have to take decisions as theorized by Lewontin (2002).

There are two interrelated issues in the confrontation between expert knowledge and social and political action. First, how are we to go about acquiring socially relevant knowledge? While there remain a few vestiges of the belief that knowledge of some aspects of material nature can come from divine revelation, these do not generally impinge on the interactions between the believers and the physical world. Apparently even the most devoted adherent of fundamentalist faith agrees that one must go to flight school to learn how to operate an airplane. Since the activities of research and development that produce scientific knowledge and its technological applications require the expenditure of a good deal of time and money, hard decisions have to be made. Should 4 billion be spent from federal funds on the Superconducting Super Collider, an atom smasher whose scientific purpose was to acquire a fundamental understanding of the structure of matter? Or should 3 billion be spent on the Human Genome Project in the hope of learning what it is to be human, not to mention curing an unspecified list of diseases? Who should decide? Congress? Scientists? Congress in consultation with scientists? Which scientists? Second, once knowledge is acquired as a consequence of the first set of social decisions, how are we to introduce that knowledge into the process of making further social decisions? How can legislators, judges, juries, school boards use knowledge of which they can have only an imperfect, if not distorted, understanding to make decisions? Of course, they must ask the experts.

In practice however (Lewontin 2002).

What is baffling in the intersection between science and politics is that, with the exception of Nixon, no administration has been honest in admitting its disdain for the opinion of the scientific community and its politically appointed representatives on matters of public policy. Congress will continue to fund the NIH because everyone gets sick, suffers, and eventually dies. It is a foolish member of Congress who votes against health. But whether or not an administration decides to push an antiballistic missile program, or cut emission standards, or bury radioactive wastes in a salt mine has seldom, if ever, depended on the analysis of its appointed scientific advisers.

The same thing will probably happen for the climate change aggravated by the fact that negationists are gaining ground in the populist government. On 5 April 2011, the *Wall Street Journal* published and essay "How scientific is climate science" completed with a technical supplement written by Douglas Keenan. Keenan defines himself *an independent mathematical scientist.* The thesis of the paper was that statistics used in the IPCC report was wrong. The paper produced a parliamentary question in November 2012 which asked *ask Her Majesty's Government... whether they consider a rise in global temperature of 0.8°C since 1880 to be significant.* The answer which involved up to the Met Office (that wrote a paper on the question) is a discouraging proof of how low and unrespected is the climate science. No one would have dared to cast doubt on the discovery of gravitational waves or the Higgs boson because such sophisticated studies cannot be approached with the knowledge of a common taxpayer. On the other hand, if the taxpayer has enough knowledge of statistical methods (any bank clerk can) he can question the IPCC conclusions.

This episode on one side confirms the consideration of Richard Lewontin *Even within the ranks of "scientists" only a tiny subset has the necessary expertise to make an informed decision about a particular issue.* while on the other hand confirms that the method used in the IPCC reports is just too simple and not supported by a sophisticated experimental proofs. This is the same reason why the decisions on important environmental issue are taken by the politicians: the scientific "elite" in climate science has not enough elitist stature. Many are the origins of such dubious scientific authority of the climate scientists but these have not been handled at all by the philosophers (the most recent) of the climate.

The main problem of the climate research is that today it is mainly identified with the research on global warming and the reason is that this is the subject which attracts most of the available funds. However, global warming should be just one of the many important subjects that confronts climate science. Actually, very few of them have been solved satisfactorily. As we mentioned in the early part of this book at this time, there does not exist a complete theory of the ice ages. It is not yet clear why the dominant period has changed before 1 million year ago. It is not yet clear why the concentration of the greenhouse gases like carbon dioxide and methane has changed coherently with the ice volume. It still obscure how the Younger Dryas episode was generated and above all there is not a complete simulation of an ice age starting from the solar forcing.

All the question of the snowball Earth is still quite obscure because the geological data are still limited and there is not a theory about the starting and the termination of the ice covered Earth. Phenomena like the North Atlantic Oscillation or the Arctic Oscillation are not explained yet. We could continue for a while but these examples are enough to show that there are plenty of problems to be studied which could have a direct relevance for society. The study of ENSO (El Nino Southern Oscillation) is a good example in this direction. Since the seminal paper by Zebiak and Cane (1987), forecasts of El Nino have been demonstrated (Chen et al. 2004). In recent years, US government agencies have funded an impressive program to forecast months in advance El Nino. The program is also based on the pioneering work of Zebiak and Cane. This remains as an example that climatic problems must be tacked with appropriate means and as in any other scientific problems experimental data are essential to solve the problem.

The scientific climate community today is concentrated mainly on the global warming which is discussed almost exclusively using GCM results. We have discussed the necessity to test the predictive capability of GCM and the only way to accomplish this is to use hindcast. At the moment, there are several data sets of assimilated data that could be used for this purpose and there are a few examples on their use (see for example Mote et al. 2016, Ou et al. 2013). The IPCC also reports these simulations in terms of average quantities and however the main problems of these studies is the large systematic errors which affect the simulation. Tredger (2009) and (Frigg et al 2015) reports error up to 3.5 °C in temperature over a trend of less 1 °C over 100 years. To overcome such difficulty official document usually report deviates from the mean. We will discuss this critical point in a few chapters however it is relavant the observation by Palmer (2016).

In the view of their relevance in solving urgent real-world problems, it is a matter of concern that climate models continue to exhibit pervasive and systematic errors when compared with observations. These errors can be as large as the signals predicted to arise from anthropogenic greenhouse gas emissions. Because the is profoundly nonlinear, these errors cannot be simply scaled away, or reliably subtracted out a posteriori with some empirical correction based on past observations, when simulating the future climatic response to human carbon emissions.

Even considering that the use of GCM is just an engineering exercise, it cannot be accepted that such large errors are neglected. This is not to deny the reality of global warning but it is rather to suggest that the global and regional details of the global warming are not realistic: they could be better or worst.

The problem is dramatically demonstrated in a work by Richard Seager and his coworkers (Seager et al. 2008) where they have tried to simulate the drought that hit the US in 1930s. An ensemble run utilized climatogical Sea Surface temperature (SST) and was compared with a GCM integration which run from 1929 to the next 10 years. The comparison showed that the climatological run reproduced with enough accuracy the drought while the model run did miss the geographical region and the severity of the drought. This exercise confirms that detailed predictive capabilities of GCM are limited.

Related to this aspect there is the precision with which the data are reported. The last IPCC document even the summary for policymakers (SPM) reports predicted changes down to the second decimal digit. Especially questionable are the radiative forcing figures. On page 11 of the (SPM) you reads *The total anthropogenic RF for 2011 relative to 1750 is 2.29 [1.13 to 3.33]* where the 2.29 data is quite ridiculous as well as the comparison with 1750. The listed error does not justify the data because nobody has ever verified the consistency of calculations on radiative forcing with measurements. As we will detail in the next chapter spectra of active radiative gases have been measured almost three decades apart but because they were not calibrated they could not be used to calculate the radiative forcing. The radiative forcing is a fundamental quantity in the simulation of global warming and does not have corresponding experimental values. Again this is a consequence of the very low priority given to measurements in the global warming research that brought to the cancelation of the CLARREO program.

All the previous considerations do not want to deny the existence of the global warming. Rather it is time to recognize that GCMs are useless to give detail of the climate change and that they result with such large systematic errors show that there must be some problem in their operation. As we have seen from the previous chapters simpler models should be used and adaptation policies should be developed. Financial resources should be devoted to more general climate studies. However, when you have armies of modelers it is very hard to dismount them.

The way the global warming problem is handled does not qualify the climate sciences to be in the realm of the physical sciences. Very important variables do not have an experimental number and do not satisfy the most elementary rules.

A last example may reinforce this view. In the philosophy of science observations are said to be "theory-laden" when they are affected by the assumption of the investigator. We will report a classical example of this situation in the next chapter. Edwards (1999) noticed that GCM are 'data-laden' meaning that they contain many "semi-empirical" parameters deriving from the necessity to parameterize processes. This point of view is symmetric with respect to the concept that observations are theory-laden. This concept is nothing strange because theories are always full of constants derived from measurements. The difference again between the physical sciences theories and the climate sciences is that the former explains large-scale effects in terms of smaller scale processes while the second never starts from first principle to build complicated GCM's. As Palmer (2016) affirms

Of course, it is far from easy to simulate global mean temperature from first principles, depending as it does on cloud cover, and hence the hydrological cycle. (It is of course possible to tune a model to have the right surface temperature. Whether the fact that this has not happened is due to scrupulous honesty on the modelers' part, or whether the models have been tuned to get some other aspect of the simulated climate correct, such as radiative balance, is a moot point.)

Climate sciences are something different from physical sciences but still should be respected. Besides as affirmed by Richard Goody "we should separate the climate policy from the climate science. The former will use GCMs and make such interpretations as they wish. The latter should be asking why the climate is what it is, using heuristic models and scientific reasoning to look at one link at a time. If GCMs are used they should be used in an heuristic (i.e., experimental) mode".

But what it is the climate science

A climate scientist and a humanist at the end of each chapter will discuss what we intended to say and what they have understood. The humanist also in this case is more familiar with the many philosophical implications.

H: This chapter stimulate me a lot because I am familiar with many of the concepts expressed. I would like to start with the sentence attributed to Charlie Dunbar Broad, induction is the glory of science and the scandal for the philosophy. I hope this does not mean that philosophy is useless.

C: Not at all! As we have said philosophy deal with absolute concept and the "science" (I put it between quotes) and/or the scientists have learned from the philosopher important lesson like the precision of reasoning, the skepticism, the necessity to formulate clearly the questions and other matter in which philosophers excels. Naturally scientists and philosophers have important differences especially when they deal with practical questions.

H: It is very hard to accept such a simple justification because I understand that SSK is a movement that questions the fundaments and motivation of science or rather that the research is influenced by the social environment in which is developed. This seems to fit perfectly for the climate sciences to the point that could be taken as an example of science as social construct.

C: We need to distinguish here what are known as climate science s and what actually should be and that we are trying to understand. Here we consider all the recent up and downs and ironically after the award of peace Nobel prize to IPCC this kind of research may represent, bluntly, a corrupt and corruptible science. It is inconceivable that today consensus should arrive from a political organ like the IPCC whose job should be to assist the governments in the preparation of the energetic policies and development and nothing else. To this end there are also United Nation organisms like UNEP. For other emergencies like AIDS there have been commissions nominated which do not include all the government representative. Just talking about these problems we are in the middle of a political problem without mentioning things like climategate. This derives from the misunderstanding that the scientists have not resolved on the real nature of their research. A good part of the results of this research are poor, just imitation of other work, which utilize more or less simple parameterization without more ambitious targets.

H: It looks like following the logic of the SSK the results are quite poor.

C: This is only true in part. There are other disciplines which maintain their ambitions that works at extremely competitive levels. I have mentioned on purpose the Nobel prize (only because it is an award of the establishment) which has been assigned for the PEACE non for any science. Medicine also receive Nobel prizes and at the same time there are humanitary organizations (like Medicins Sans Frontieres) related to the medical profession that have received it while in this case the award has stopped to the Al Gore stuff.
The SSK thinking may be right and it is something to keep in the right social perspective but it is not influencing the research. In this case is the political power which corrupt the climate science s through the nomination of the experts in the IPCC.

H: It is peculiar the influence that the GCM results have on vast social sectors from the economy to the energetic policies or the mitigation programs. Al these are based on the models.

C: This is the reality of the present research and it is related to the quality we mentioned before. The predictions for the purposes you just mentioned could be made through the use of simpler models that not require large investement and especially the dependence from a few research centers. The present knowledge is such that does not allow to study the regional effects which in most cases are obtained with a simple telescopic effect without adding any information. However the situation is such that it will be very hard to convice the scientist to a have a reasonable behavior. On the other hand it is necessary that the climate studies be based on a more rigorous approach.

H: I understand that more than physical science, statistics could be relevant. In particular I am intrigued by the example on the true nature of the coin.

C: The example needs to establish the difference between a deductive and inductive reasoning. In the deductive case the simplest example is if A=B and B=C, then A=C. The conclusion was already in the premises. On the contrary in the induction from a single fact you arrive to a general conclusion. In the case of the coin you can apply the statistical methods to determine whether it is fair or not. That is you use a model to arrive at the data. In the case of the induction your start from the data and obtain a model utilizing the statistics. You can look for example to the statistics of the tosses. It is evident that in the case of the climate and inductive process is more appropriate. An example can be found in the evolution of the numerical weather predictions (NWP). Before the advent of the NWP there was an analysis of the weather situation and based on that a qualitative forecast could be made. This was a complete deductive approach. The present NWP give a total detail on the forecast so that the prediction has lost it deductive character. But there is more than that. The weather forecast can be tested day by day while this cannot be done for the climate prediction. However these can be improved as new data become available in the years following the first predictiom. That is the model (or whatever it is) can be corrected as new data become available. There is statistics that was invented more the two century ago that can do that and is the Bayes statistics.
H: The conclusion bring us inevitably to the data and that these are the last object that confronts the validity of the theories. I do not understand the Lewontin outburst.
C: Sincerely it is a little annoying the tone of the scientists as exponents of the establishment and in this sense the Gauch sentence about "finding the truth" may be exaggerated. This is the reason why Lewontin is important.
H: The Harod Jeffreys opinion on the "inference from empirical data" is then extremely aseptic and apt. However the opinion of Naomi Oreskes remain on the impossibility to validate the models.
C: In general there is nothing new when we talk about the comparison with data but we need to understand the nature and quality of the data. The objections raised to the validation of the models is another proof that the approach must be changed for the climate prediction or at the least to invent new procedures that make them more reliable. This can be summarized with the statement that the prediction must be nearest to the evidence. This sentence we will try to explain in the following chapter.
H: The last paragraph discussion of the nature of climate sciences looks quite interesting and also a little radical. The main point is the identifications of the global warming problem with the "climate problem" which is not. Also it pinpoints all the contradictions in the use of GCM.

C: The last paragraph gives a rather accurate summary of the present situation. However the critiques are not directed to deny the existence of global warming but rather to the methods used to study it. It is impressive that the radiative forcing has never been measured experimentally and for decades the community has denied any experiment to correct this problem. There are though universities and other research institutions which pursue climate programs involving chemist or paleontologist. From this point of view the study of climate looks more like medicine that utilizes different sciences to develop cures for the patients. Exclusive trust in GCM's would corresponds to use aspirin for all kind of illness.

References

Chen, D., Cane, M. A., Kaplan, A., Zebiak, S. E., & Huang, D. J. (2004). Predictability of El Nino over the past 148 years. *Nature*, *428*, 733–736.

Edwards, P. N. (1999). Global climate science, uncertainty and politics: Data laden models, model filtered data. *Science as Culture*, *8*, 437–472.

Frigg, R., Thompson, E., & Werndl, C. (2015). Philosophy of climate science part II: modelling climate change. *Philosophy Compass*, *10*(12), 965–977.

Gauch, H. G. (2002). *Scientific methods in practice*. Cambridge: Cambridge University Press. Reprinted by permission from Cambridge University Press.

Jeffreys, H. (2010). *Scientific inference*. Cambridge: Cambridge University Press. Reprinted by permission from Cambridge University Press.

Lewontin, R. C. (2002). The politics of science. *New York Review Books*, *49*(May 9). from The New York Review of Books Copyright © 2002 by Richard C. Lewontin.

Lewontin, R. C. (1990). Fallen angels. *New York Review Books*, *37*(June 10) issue from The New York Review of Books Copyright © 1990 by Richard C. Lewontin.

Lewontin, R.C., Rose, S. & Kamin, L. J. (1985). Not in our genes: Biology, ideology, and human nature. Random House permission from the authors.

Mote, P. W., Allen, M. R., Jones, R. G., Li, S., Mera, R., Rupp, D. E., et al. (2016). Superensemble regional climate modeling for the western United States. *Bulletin of the American Meteorological Society*, *97*, 203–216.

Ou, T., Chen, D., Linderholm, H. W., & Jeong, J. (2013). Evaluation of global climate models in simulating extreme precipitation in China. *Tellus A*, *65*, 19799.

Palmer, T. N. (2016). A personal perspective on modelling the. *Proceeding of the Royal Society A*, *472*, 20150772.

Pickering, A. (1984). *Costructing quarks: A sociological history of particle physics*. Chicago: Chicago University Press.

Seager, R., Kushnir, Y., Ting, M., Cane, M., Naik, N., & Miller, J. (2008). Would advance knowledge of 1930s SSTs have allowed prediction of the Dust Bowl drought? *Journal of Climate*, *38*, 3261–3281.

Shackley, S., Young, P. C., Parkinson, S., & Wynne, B. E. (1998). Uncertainty, complexity and concepts of good science in climate change modelling : Are GCMs the best tools? *Climatic Change*, *31*, 159–205.

Shackley, S., Risbey, J., Stone, S. P., & Wynne, B. E. (1999). Adjusting to policy expectations in climate change modeling - An interdisciplinary study of flux adjustments in coupled atmosphere-ocean General Circulation models. *Climatic Change*, *43*, 413.

Tredger, E. (2009). *On the evaluation of uncertainties in climate models*, Ph.D. thesis, London School of Economics

Zebiak, S., & Cane, M. A. (1987). A model El Nino-Southern oscillation. *Monthly Weather Review*, *115*, 2262–2278.

Chapter 5
Experimental Data and Climate

5.1 Introduction

Physical sciences can be grouped in two categories: the experimental sciences (chemistry and physics) based on the simplest form of evidence, the controlled experiment, and the observational sciences (earth, atmosphere, oceanic sciences, and astronomy) whose evidence is made up of observations of immense systems that cannot be controlled experimentally. In both cases, however, the approach is based on the experimental proof to eliminate unsatisfactory theories. In any case, the approach of the single researcher is different if he has to deal with a controlled experiment or with an immense system he cannot control.

Climate prediction is part of atmospheric and oceanic sciences (that is observational) that have further complication related to the use of powerful computers necessary for the modeling. An environmental model tries to couple all physical and chemical components of a natural system (the climate) in a framework based on the equations that regulate the motions of a fluid. The models are subjected to external conditions that change (forcings) and projects to the future, the present conditions through a series of small time steps, that are the numerical integration. Environmental models may have a considerable complexity and could produce very detailed predictions. As we mentioned before, Keith Beven has published a book in 2009 on the modeling of the environment where he suggests that models have introduced a new point of view in science that was not known before and that he calls *pragmatic realism*. This point of view regards the output of the environmental model as a form of reality that can bring to the "ultimate" reality by improving the model's component. This concept is based on a dubious vision of reality and constitutes a further step away from the experimental conception of the physical sciences.

We have discussed how the climate predictions are made by environmental models known as General Circulation Models (GCM) and how their results imply considerable uncertainties. The same uncertainties are the main argument of the critics of the climate change and are the main reason why progress in the climate regulations have been so slow. Without any doubt, the political opposition that is still registered

G. Visconti, *Problems, Philosophy and Politics of Climate Science*,
Springer Climate, https://doi.org/10.1007/978-3-319-65669-4_5

today would have no basis if the uncertainties on the climate predictions would be drastically reduced. This should be the purpose and aspiration of the climate community.

In the recent history, the successes of the experimental sciences have been outstanding from the control of nuclear energy to the reading of the genetic code, and these results are at the base of the respect that science deserve from society. It is also true that the systems of the experimental scientist are basically simpler and the uncertainties can be controlled. Their objective is a precision of measurement that could distinguish between competitive hypothesis. They also can change a few parameters of the experiment, to produce a large amount of data. This chapter asks if there is a lesson to learn from the experimental sciences that could be used for an improvement in the quality of the climate predictions. It could be that the uncertainties in the predictions will not be eliminated, but this could be a step in the right direction. The link could be a better use of the experimental proof and in this sense the difference is that for the experimental scientist the evidence dominates all the other considerations while for the climate prediction the emphasis seem to be the pragmatic realism.

5.2 Uncertainties in the Climate Prediction

The subject of climate change and the climatic predictions is now defined by the periodic report of the IPCC (Intergovernmental Panel on Climate Change). In the last edition, hundreds of scientist from dozens of countries report the consensus of almost the totality of the climatic community. One of the results of such a document is the simulation of the global mean temperature of the Earth, as shown in Fig. 5.1. This is not exactly obtained by IPCC document but the figure reports 38 predictions of the global mean temperature in the presence of the forcing from the greenhouse gases from the past century. The difference with the IPCC document is that, here the temperatures are predicted by the model. In the official documents, the anomalies of different models with respect to some average are reported. The figure shows a general increase of temperature for all models with a spreading of something like 2.5° that is a difference so large to influence the debate on the climatic change. Notice that these simulations refer to a historical period (the last century) and the spread in large measure is caused by the difference between different models but also to the year-to-year fluctuations. These may be caused by nonlinear instabilities in the models and correspond somehow to the variability observed in the "natural" climate and that correspond to the weather (see the discussion on the macroweather). Because these simulations are historical, other causes for differences are not included like the economical and political factors that may influence the predictions of the future climate.

CMIP stands for Climate Model Intercomparison Program and as it reported in the corresponding website

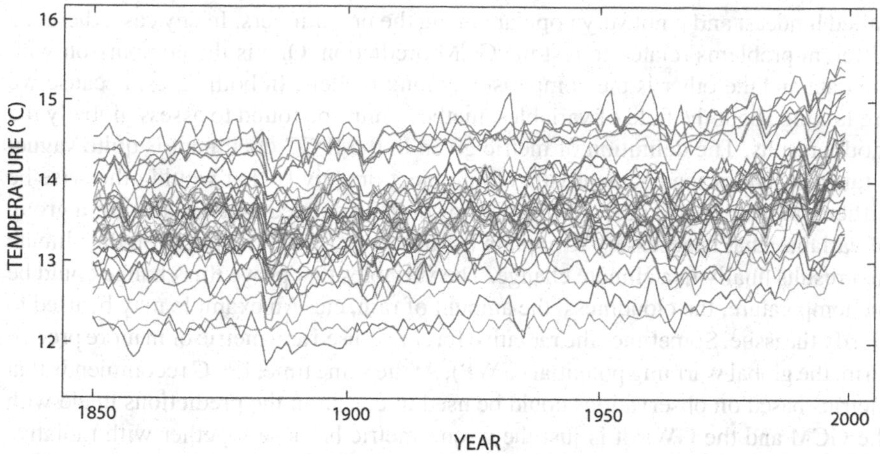

Fig. 5.1 Model global mean temperature over the period 1850–2000 for the CMIP 5 ensemble. Different colors refer to different models. (Guttorp 2014)

At a September 2008 meeting involving 20 climate modeling groups from around the world, the WCRP's Working Group on Coupled Modelling (WGCM), with input from the IGBP AIMES project, agreed to promote a new set of coordinated climate model experiments. These experiments comprise the fifth phase of the Coupled Model Intercomparison Project (CMIP5). CMIP5 will notably provide a multi-model context for (1) assessing the mechanisms responsible for model differences in poorly understood feedbacks associated with the carbon cycle and with clouds, (2) examining climate "predictability" and exploring the ability of models to predict climate on decadal time scales, and, more generally, (3) determining why similarly forced models produce a range of responses.

Greater details on the CMIP 5 could be found in Taylor et al. (2012).

The suspicion is that if you present such results to any meeting of physicists, they will point out that something should be wrong with the models because of the large spreading and because there is a difference of about one degree between the average prediction and the data. It is rather doubtful that models that show such large spreading could be trusted when they predict a change much less than their spreading. A very crude view is that at the least the systematic error is large. Similar representations could be found in Frigg et al. (2015) or (Tredger 2009).

There is a chapter in the IPCC report on model evaluation where the so-called climate metric is introduced. Metric is one of those concepts that seems to be introduced to moderate the inferiority complex that climate modelers have with respect to scientists from other disciplines. As it is customary in physics, theories are compared with experimental data. It must be stated that GCMs do not represent the expression of any theory on how they works. Models are "engineering" constructions used mainly to simulate past, present, and future climate. In the last application, the appropriate word is not forecast, but rather prediction. The simulation of past (recent) climate is called hindcast and it not very popular among the practitioners. In any case, there are different problems related to test the GCM prediction. One is the comparison with the data and the other is the comparison among models. In both cases, because we are talking about the field of variables, methods must be found to assess globally the model results. The definition of metric given in the IPCC document is quite vague: a fair measure of the characteristic of an object or activity that would otherwise be difficult to measure. In other terms, metric could be defined as a variable (or a group of variable) that is necessary to explain the influence of some process on the climate that results finally in a climate change. The variables that prove this change could be the temperature, the cloudiness, the amount of rain, etc An example may be used to clarify the issue. Sometimes the radiative forcing is used as a metric or in more precise term, the global warming potential (GWP). At the same time, IPCC recommends that metrics based on observations could be used to constrain the predictions made with the GCM and the GWP it is just the wrong metric because together with radiative forcing is a quantity that at the present time has not been verified experimentally . This is probably one of the most critical aspect of the climate prediction that requires urgent attention. This especially considering that in the future we may expect several ensemble predictions that need an appropriate metric to select the most probable. We will examine some question related to the development of metrics that could be used for the model evaluation and what are the most useful data in this respect. All the atmospheric variables could be used for this purpose or even their combination, so there are several observations that could be the basis for a metric. But at the present time, very few of them have been used.

There is however another point of view which is referred as "truth plus noise" (Massonnet et al. (2016), Siegert et al. (2016)) or "truth plus error" (Sanderson and Knutti 2012). Actually Siegert et al. and Sanderson and Knutti deal with the interpretation of ensemble simulation. In particular, the latter refers to two different integration of the ensemble. In one case, each model of the ensemble is an approximation of the true system with some random errors. In the other case, the true system can be interpreted as a sample drawn from a distribution of models, such that models and truth are statistically indistinguishable. The discussion in this case is centered on the fact whether the spread on the historical simulations can be used to estimate the difference in the simulations on the future. The answer apparently is negative, because the future simulations are dominated by the different treatment of physical processes and feedback mechanisms.

On the other hand, the approach of Massonet et al. is much more radical, because it starts from the assumption of total symmetry between models and data. Models represent the true status of the system and some errors while observation also are affected by uncertainties. The conclusion is that if the metric used to evaluate models is a measure of "quality" of the model performance, then the same metric can be used to assess the quality of the observation. The trouble is that at the end, the paper must use data from El Nino to assess the quality of the forecast. In any case, the quality of the data is an important factor in the evaluation of the climate processes.

5.3 The Quality of the Climate Data

As shown in Fig. 5.1 the predicted change in temperature is about 1° per century. In order to discriminate among different theories the measurement should have a precision of at least 0.1° and this precision should be maintained at the least for a century. The meteorological data do not have this kind of precision even when they are reanalysed. This means that all the available data are made consistent using a numerical weather forecast model. The required precision can only be obtained through a systematic calibration of the measuring device. This calibration must be done directly with international standards and this is what is called a benchmark. Paradoxically while it is rather complicated to make a benchmark on a radiosonde it is relatively simple (in principle) to do the same on a spaceborne instrument.

At the present time, there are two kinds of measurements that use an orbit calibration. The first experiment measures the solar radiation and it is called ACRIM (Active Cavity Radiometer Irradiance Monitor) that utilizes a sophisticated calibration system to measure the radiation emitted by the Sun. This instrument takes measurements with a precision of 0.01%. Thanks to this instrument, the solar radiation has been measured over three solar cycles (about 30 years) since the beginning of 1970 with three different spacecrafts. The connection between the three periods has been possible because the measurements were calibrated.

The second measure which is directly comparable with the international standards refer to the radio occultations methods. This data is a byproduct of a very sophisticated measurement that now is used by anyone and it is known as Global Positioning System (GPS). The exact position of your car is obtained by a triangulation method based on the time of arrival of the signal from three different satellites to the GPS receiver in the car. The reference clocks on the satellite are calibrated with atomic clocks and then can be referred to international standards. Meteorology enters in the GPS algorithm, because the radio signal that goes through the atmosphere is influenced by the refraction: the same phenomena experienced by a light ray when passing through the water. In practice, the radio signal that goes through the atmosphere is curved and this deformation implies a time delay from which the refraction index of the air can be obtained. From the refraction index, it is possible to obtain pressure and temperature of the region traversed by the signal as well as its water vapor content. It is rather interesting that the first applications of this method have been made with the signals of probes that were exploring other planets like Venus and Mars. This radio occultation method can also be referred to international standards, because it is based on the measurement of time. Using this method, about 20 years of data on the distribution of temperature and pressure have been accumulated.

A third method that has never been utilized in orbit deals with an instrument calibrated in terms of temperature standards. We need to make a slight detour to introduce how the solar radiation and the infrared radiation emitted by the Earth interacts with the climatic system. Figure 5.2 illustrates the distribution of the infrared radiation emitted by the Earth as a function of the wavelength. This is the so-called spectrum of radiation and when examined carefully could give very useful information.

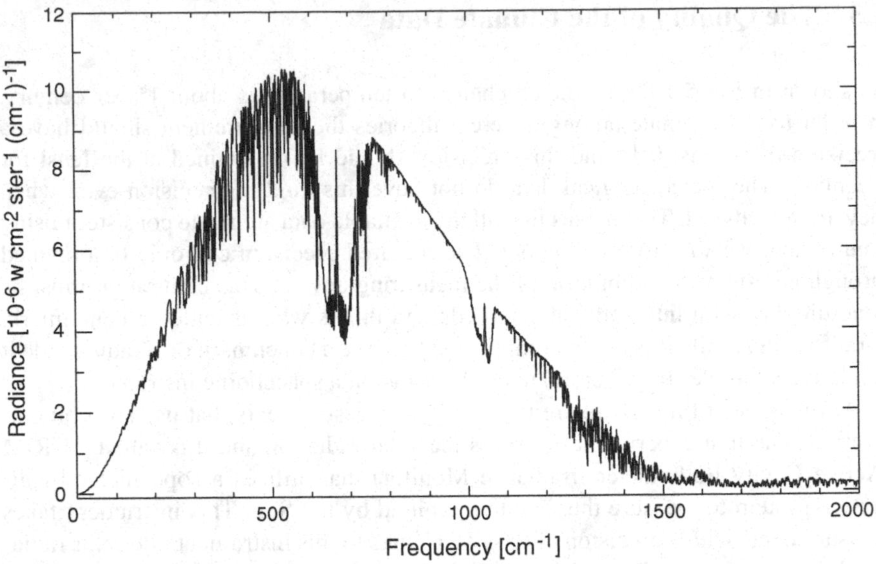

Fig. 5.2 The spectra of the Earth emitted infrared radiation. Units of wavelength are in cm^{-1}, To obtain the wavelength in micron just make the inverse and multiply by 10,000. The CO_2 15 μm band correspond to roughly 667 cm^{-1} while the ozone 9.6 μm corresponds to 1041 cm^{-1}. (Goody and Visconti 2013)

The area subtended by the spectrum is related to the power emitted by the planet and at the equilibrium is exactly the same as that absorbed by the Sun. This is one of the basic relationships that establish the temperature of the Earth and which depends on the irregularities that marks the spectrum. Beside the many "wiggles" there are two large "holes" around the wavelength of 15 μm (667 cm^{-1}) and around 9.6 μm (1041 cm^{-1}). These two signatures are attributed to the carbon dioxide and ozone absorption. All the other irregularities are due to the presence of other gases found in the atmosphere. In Fig. 5.3, the presence of such gases is very much in evidence and it is obtained examining two spectra taken 27 years apart (between 1970 and 1997) by two different instruments aboard the satellites: (IRIS, Infrared Interferometric Spectrometer) aboard Nimbus 4 and the more recent IMG(Interferometric Monitoring of greenhouse gases) on the Japanese satellite ADEOS (Advanced Earth Observing satellite). The spectra are taken both on the region of Central Pacific and are plotted in a different fashion with respect to Fig. 5.2. In this case, rather than the power, the spectra is plotted as "brightness temperature" as a function of wavenumber (the inverse of the wavelength). Brightness temperature gives a direct indication of the region from which the radiation originates.

Most of the spectra show a temperature around 290 degree Kelvin (K) which is clearly the sea surface temperature. On the other hand the CO_2 band goes down to about 240 K indicating that originates around the tropopause while the O_3 band originates in the stratosphere where the temperatures are a little higher . The absorption

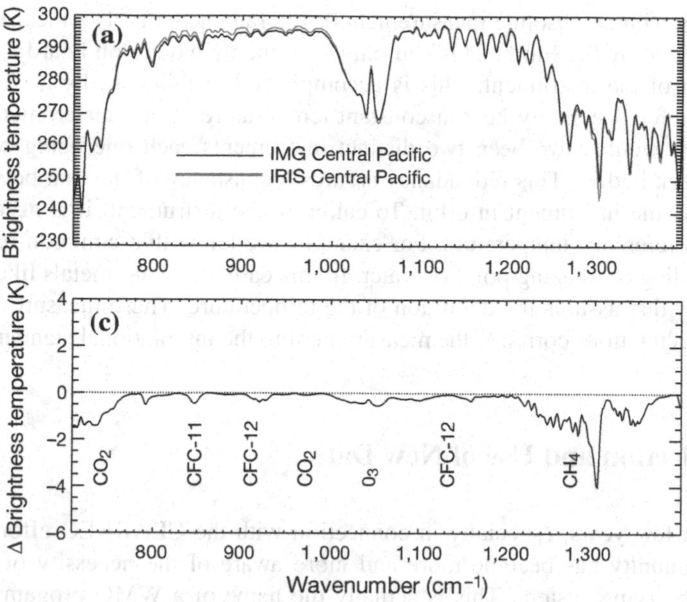

Fig. 5.3 The spectra of the emitted infrared radiation from Earth taken 27 years apart. The figure at the *top* shows the brighntess temperature for the two spectra while the figure at the *bottom* shows the contribution of each of the greenhouse gases. The figure is adapted from the paper by (Harries et al. 2001). See text for a more detailed explanation

of water vapor is responsible for the lower temperatures between 1200 and 1400 cm^{-1} while distinct absorption feature due to methane (CH_4) is evident around 7.6 or 1300 cm^{-1}.

The interpretation of these data is not simple just because the data are not calibrated and the comparison is very much qualitative, but this kind of data may be helpful to understand the greenhouse effect. If the amount of greenhouse gases increases, the absorption due to the same gases increases and consequently also the areas of the corresponding "holes" in the spectra. However, if equilibrium must be maintained between absorbed solar radiation and emitted radiation, then the area subtended by the spectra must be constant because it is constant during the solar radiation. The only way to compensate for the larger holes is the increase of the surface temperature. This is the mechanism of the greenhouse effect: if the amount of greenhouse gases increases, the absorption increases as the temperature.

A spectra like the one in Fig. 5.2 is registered in space by an instrument called interferometer. It has been invented by a NASA scientist, Rudolf Hanel that built it for the Mars probe Mariner 4 back in 1964. The instrument is based on the one used by Michelson and Morely to demonstrate the nonexistence of the ether. After being used to study Mars, the same instrument was put on board of a weather satellite of the Nimbus series. A much recent version of this instrument was planned to be on one of the satellite of the CLARREO project. As specified by the acronym, we are dealing with an absolute measurement of the radiance, that is a measure of the power emitted by the Earth. This can only be done by having on board a constant calibration of the instrument. This is accomplished in this case by a black body source, a reference cavity kept at constant temperature. The interferometer of the CLARREO would have been two distinct instruments each one being calibrated by two black bodies. This redundancy assures a constancy of the link between the ground and the instrument in orbit. To calibrate the instrument, it is necessary to measure a reference temperature. For example, we know that common references are the boiling or freezing point of water. In this case, there are metals like gallium or mercury that assures the definition of the temperature. These measures as those of radio occultations correlate the measurement to the international standard.

5.4 Selection and Use of New Data

In the last few years, especially in connection with the CLARREO effort the climate community has become more and more aware of the necessity of a global climate observing system. This is actually the name of a WMO progam (GCOS) whose *goal is to provide comprehensive information on the total, involving a multidisciplinary range of physical, chemical and biological properties, and atmospheric, oceanic, hydrological, cryospheric and terrestrial processes*. On GCOS site you also read *GCOS is intended to meet the full range of national and international requirements for climate and climate-related observations. As a system of climate-relevant observing systems, it constitutes, in aggregate, the climate observing component of the Global Earth Observation System of Systems (GEOSS)*. Being an intergovernmental organization, GCOS has somewhat limited capability, so that the community is rather moving in the direction of reproducing for the climate what in the past has been done for weather forecasting with the so called Observing Systems Simulation Experiments (OSSE) and which is described for example in (Arnold and Dey 1986). The experiments to which the acronym refer are actually sensitivity studies carried out with a computer simulation and are contrasted to the real data experiments. Simulations are mainly used as evaluations of proposed remote sensing system. It was quite natural that the same concept would be extended to climate but in this case the huge database of weather observations is missing. The new system was reported as COSSE, Climate Observing Systems Simulation Experiments. One of the principal advocates of COSSE is Bruce Wielicki of NASA Langley and he indicated the main objectives of COSSE as process observations besides long-term change observation while the observing system would include satellite, surface, in-situ and field experiment observation. The emphasis in designing COSSE is mostly on practical and achievable objectives. In one of the presentation of Bruce Wielicki this approach is summarized as "investigators seem to have settled for what is measurable instead of measuring what they really like to know".

Based on that, COSSE studies reported some quantification of the science objectives which included for example, narrow the uncertainty in climate sensitivity by a factor of 2; determine the rate of change of sea-level rise to 0.2 mm/yr; measure the tropic expansion with a precision of 15 Km/decade; determine ocean heat storage to within 0.1 w/m^2 for 10 year average or 0.2 w/m^2 for annual average; determine Greenland and Antarctic ice sheet mass loss to within 0.1 mm/yr of sea level rise equivalent (that is how much ice loss produce the indicated sea-level rise). Some of the objectives are based on the CLARREO requirements. We have illustrated some of these effects already, except for the tropic expansion. Associated with global warming, an expansion of the tropical Hadley cell is expected with the consequence that the high pressure associated to the descending branch on the cell will affect the precipitation in that region negatively.

It is interesting to establish a correspondence between OSSE and COSSE. While for OSSE, the initial conditions for models are essential for COSSE is the decadal tendency. The climate processes to be studied by COSSE could range from initial conditions to ocean deep circulation. Similar to what happens for OSSE, COSSE should enable better communication/coordination of climate observations and climate modeling research communities. An example of how COSSE could work is given by the study of the sea-level processes. In this case, the relevant measurements are gravity (for the ice loss), altimeter (for the sea level) and the reference frame (to interpret the data). The objectives are to establish sea level to an absolute value of 1 mm/decade, ocean heat storage within 0.1 w/m^2 per decade. These determinations will directly imply application to the economic impact of shoreline inundation.

The economy of a COSSE has been discussed by (Cooke et al. 2014, 2016) and the investment is estimated to be in the range of 10 trillion dollars and this would advance the climate science by 15 years, producing a return of \$ 50 for every \$ 1 invested by society. This approach is very much different with the present struggle to get some funding for climate research project.

Interesting as it is that COSSE initiative did not get any attention from NASA or other agencies and universities and only the Jet Propulsion Laboratory in Pasadena is implementing COSSE for what concerns the sea-level problem.

We have mentioned several times CLARREO and it is interesting to report on some of the potentialities of this now defunct project especially in connection with the problem of establishing which are the best data that could help us to study the global warming . This topic would be completed in the next chapter when we discuss Bayes statistics in detail. We would like to mention just two topics: the first one deals with the detection of a temperature trend and the other one is about the determination of the feedback factors.

As for the temperature trend if we refer to the IPCC estimate of 0.2 K per decade it is possible to show that even for a *perfect observing system*, a climate record of 12 years is necessary for a 95% confidence level while to detect a 0.1 K per decade trend the same system require 22 years. This reaffirms the necessity of a constant calibration for a space observing system. The sources of uncertainties include natural variability, absolute calibration uncertainty, instrument noise, and orbit sampling uncertainty. The calibration accuracy has dramatic effect on the time of detection. CLARREO has a calibration accuracy of 0.06 K while the current weather satellite have accuracy between 0.24–0.36 K and this means that while for CLARREO a little more than 10 years is necessary to detect a 0.2/decade trend a weather satellite will take more than 30 years . The same consideration could be made to detect changes in the cloud radiative forcing but with much longer detection times. Most of these data are taken from the paper by (Wielicki et al. 2013).

A practical example of COSSE could be made with the studies carried out for the determination of the feedback parameters using CLARREO measurements. We have mentioned in one of early paragraphs (1.2) that the climate sensitivity increases with the feedback processes. Now we can do something better and in the simplest case (absence of feedback) the sensitivity λ_0 it is simply

$$\lambda_0 = \frac{\Delta F_{rad}}{\Delta T}$$

where ΔF_{rad} is the change in radiative forcing. If we add a generic feedback with sensitivity λ_i, the corresponding change in radiation is $\Delta F'_{rad} = \lambda_i \Delta T$ and the cumulative sensitivity would be

$$\Lambda = \lambda_0 - \sum \lambda_i$$

and a magnification factor can be defined as

$$g = \frac{\lambda_0}{\Lambda} = \frac{\lambda_0}{\lambda_0 - \sum \lambda_i}$$

then if we rearrange the relation between radiative flux and temperature change, we get

$$\Delta T = \Delta F_{rad} \Lambda = \Delta F_{rad} (\lambda_0 - \sum \lambda_i)$$

Each feedback term can be splitted in long wave feedback and short wave (solar) feedback, and for a generic feedback it can assumed that

$$\lambda_i = (\textit{flux change/mixing ratio change})(\textit{mixing ratio change/temperature change})$$

for example, for water vapor measuring all the quantities appearing in the formula, it is possible to obtain the feedback factor. Notice that we are talking about very small values of the radiance trend of the order of 10^{-9} Wcm^{-1}(cm^{-1})$^{-1}$ster^{-1}yr^{-1} and this requires an adequate precision.

Stephen Leroy and his coworkers have explored which feedback could be studied with such method and they showed that for example, infrared spectra could be used to study water vapor feedback for the troposphere, while shortwave radiance could be used to define the cloud forcing. We will discuss some of these topics in the next chapter in the framework of Bayes statistics.

Experimental data and climate

A climate scientist and a humanist at the end of each chapter will discuss what we intended to say and what they have understood. The humanist is not very used to experimental data.

H: I understand that although the topic may appears outside the philosophical approach there are actually many papers on the use of data published on the philosophy literature.

C: You are right this is a too important topic and require a very careful and precise approach to select the type of data, how they are acquired and how they are used. Consider for example the definition of climate and climate change.

H: I was just asking that. The very intuitive definition of climate based on the average weather is not accepted any more and more sophisticated ones seem to arise. I am particularly interested to the one connected with the climate simulator

C: There is an example due to Roman Frigg of London School of Economics that may clarify this point. Suppose we chose the interval where we make average to be between times t_1 and t_2 and then just in the middle of this interval there is a catastrophic event (like a meteorite hit) that change completely the climate thereafter. We actually have two regimes completely different. This example emphasize the fact that in order to have a meaningful statistics the forcing conditions must remain constant. We actually do not know the real future forcing so the predictions are really dependent on the adopted scenario. The concept of climate simulator emphasize the role of statistics in the definition of climate. You can run your simulations with different initial condition and then using an appropriate statistics decide about the climate.

H: This call into question the statistics or more in general experts that can decide about the climate

C: As a matter of fact Rougier defines the experts as someone/something whose probabilities you accept as your own. He for example is a statistician and so he will pick up his expert based on the statistics knowledge. On the other hand with this in mind you can have your own expert based on your personal criteria. Rougier reccommend not to suppress debate within the climate community because this makes space for non scientist to be people's experts. Carl Wunsch is rather explicit on this point ...*One has the bizarre spectacle of technical discussions being carried on in the news columns of the New York Times and similar publications, not to speak of the dispiriting blog universe. In the long-term, this tabloid-like publication cannot be good for the science which developed peer review in specialized journals over many decades beginning in the 17th Century for very good reasons.*

H: Again you have discussed the quality of data but I see also problems related to the contamination between models and data.

C: The first contamination of the data by models is because data are irregularly spaced with fewer and fewer locations as we go further in the past. In this case a common technique is to use weather or climate models to interpolate or fill the missing data. This procedure correspond to the so called *reanalyses*. For paleoclimate data you do not have direct measurement but rather indirect techniques.
H: I know that there is an interesting and classical philosophical problem related to this question and it is the so called theory-ladennes of observation. The problem was discussed by Kuhn, Popper and Feyerabend
C: A very good example comes from the famous debate between John Christy and Roy Spencer of the University of Alabama and their coworkers and the climate community about the satellite measurements of global mean temperature trends. The satellite instrument did not measure directly the temperature but rather converted microwave data to temperature using a rather complex algorithm. The question raised by Christy and Spencer was that while the surface temperature showed a positive trend the tropical troposphere did not show any change. A commission of the National Research Council was appointed and produced a report that was inconclusive saying that the surface change was real while the atmospheric changes were absent. Dick Lindzen in a 2002 paper suggested a way to overcome the discrepancy. However some "philosopher" like Elisabeth Lloyd and the same Roman Frigg have insisted that the models were right while the early data showed a few important contradictions.
H: Well that could be the proof that philosophers see another side of the problem and that is the strong interaction of the social values with the climate problems. Eric Winsberg that you mentioned before has written a paper where he asserts that model-assigned probability about future climate is strongly influenced by social values.
C: I would like to turn back to the question of the quality of the data and to the ways that have been suggested to select the type of data best suited to reveal climate change. In this case the work mentioned of Huang is a very nice simulation about the value of hypothetical data taken from a spaceborn platform. The extension of this work by Stephen Leroy incidentally shows again the poor performance of models.
H: Again there is social side even on this matter. As I understand the main opposer of the CLARREO project was the meteorological community that apparently feared it had to share part of the funding with this new and important experiment.
C: Yes that could be a reason but I suspect that another reason is the fact that the government (and this is a general attitude of all the governments) do not want to obtain a definite answer to the problem of climate change. Because in that case they would be obliged to impose some regulation on the production of greenhouses gases. So the government are willing to participate to common enterprise where they have some margin for action but are not willing to take a direct initiative.

References

Arnold, C. P., & Dey, C. H. (1986). Observing-systems simulation experiments: Past present and the future. *Bulletin of the American Meteorological Society*, *67*, 687–695.

Cooke, R., Golub, A., Wielicki, B. A., Young, D. F., Mlynczak, M. G., & Baize, R. R. (2016). Using the social cost of carbon to value earth observing systems. *Climate Policy*, *17*, 1–16.

Cooke, R., Wielicki, B. A., Young, D. F., & Mlynczak, M. G. (2014). Value of information for climate observing systems. *Environment Systems and Decisions*, *34*, 98–109.

Frigg, R., Thomson, E., & Werndl, C. (2015). Philosophy of climate science Part 1: Observing climate change. *Philosophy Compass*, *10*(12), 953–964.

Goody, R. M., & Visconti, G. (2013). Climate prediction: An evidence-based perspective. *Rendiconti Lincei: SCIENZE FISICHE E NATURALI*. doi:10.1007/s12210-013-0228-2. Reproduced with permission of Springer.

Guttorp, P. (2014). Statistics and climate. *Annual Review of Statistics and Its Application*, *1*, 7–101. Reproduced with permission of Annual Review © by Annual Reviews.

Harries, J. E., Brindley, H. E., Sagoo, P. J., & Bantges, R. J. (2001). Increases in greenhouse forcing inferred from the outgoing longwave radiation spectra of the Earth in 1970 and 1997. *Nature*, *410*, 355–357. Reprinted by permission from Macmillan Publishers Ltd.

Massonnet, F., Bellprat, O., Guemas, V., & Doblas-Reyes, F. J. (2016). Using climate modekls to estimate the quality of global observational data sets. *Science*, *354*, 452–455.

Sanderson, B.M. & Knutti, R. (2012). On the interpretation of constrained climate model ensembles. *Geophysical Research Letters*, *39*, L16708. doi:10.1029/2012GL052665.

Siegert, S., Stephenson, D. B., Sansom, P. G., Scaife, A. A., Eade, R., & Arriba, A. (2016). A Bayesian framework for verification and recalibration of ensemble forecasts: How uncertain is NAO predictability? *Journal of Climate*, *28*, 995–1012.

Taylor, K. E., Stouffer, R. J., & Meehl, C. A. (2012). An overview of CMIP5 and the experiment design. *Bulletin of the American Meteorological Society*, *93*, 485–98.

Tredger, E. (2009). *On the evaluation of uncertainties in climate models*. Ph.D. thesis, London School of Economics.

Wielicki, B. A., et al. (2013). Achieving climate change absolute accuracy in orbit. *Bulletin of the American Meteorological Society*, *94*, 1519–40.

Chapter 6
The Bayes Statistics and the Climate

6.1 Introduction

Before we continue, we need to make some point on the state of our work. We have an idea on the connection between what we are attempting and Beven's book on Environmental Modelling. Beven does not try to offer solutions. Instead he asks: What is the nature of this problem? What tools are available? How can these tools best be used? The book then becomes suitable only for the graduate student (or others who wish to learn rather than pontificate), and this is the audience that needs to be consulted. As far as we can understand the problem that we have with climate prediction, it is that the "authorities" (the IPCC) have only limited objectives: to produce increasingly detailed predictions and to do so with larger and larger models. On the other hand, the climate skeptics simply point out the unquestionable weaknesses in the model predictions and ask whether an industrial revolution should be based on such flimsy evidence. This question may be resolved politically since the Environmental Movement has a great deal of support and is often not concerned about scientific integrity. This suggest that our audience could be a new generation of graduate students which is concerned with scientific integrity.

A second general question has to do with the acceptance of climate community's view that this whole discussion is about using GCM to predict the future climate. We do not believe that this is so. If you run a GCM without artificial constraints, we understand that the climate takes off to incredible states, but that all GCM's have been adjusted arbitrarily so that without industrial or solar forcing no change is predicted over a period of 1000 years. The result is that when forcing is added to the model we are only calculating the incremental effect of this forcing and not the actual climate (even if you believe that this is possible). Thus the models are being used in a heuristic mode not a predictive mode. The result is a more limited but much more plausible statement. There are certainly unrecognized climate forcings, but it is plausible that these do not affect the incremental effect of industry.

Until now we have made a long introduction on how to make the climate predictions and the possible ways to verify them. We hope that at the least one thing is

G. Visconti, *Problems, Philosophy and Politics of Climate Science*,
Springer Climate, https://doi.org/10.1007/978-3-319-65669-4_6

clear and that is, unlike the weather predictions there is no way we can verify the climate predictions, because the data necessary for that are not available yet. There is a hint of a possible method that at the least gives an estimation on the reliability of the model used. This is the so-called hindcast that is a "prediction" of past changes starting, for example, from initial conditions in the 1950s and comparing the results from the next 50 years with one of the data set which is now available. Also looking at just the last few years, there is an evident case of failure of the climate prediction and this has to do with the so-called "warming hiatus" that is the apparent slowing down of the warming trend which has characterized the average global temperature starting by the end of the 1990s.

Even if the hindcast is successful, (in the literature, there are very few simulations of this kind) still this would not prove that the predictions are correct. The only way to reach this conclusion is to "trust" the model, but then you should define possibly in a quantitative way the concept of trusting which is familiar, almost unconscious. When we board a plane, we implicity "trust" that the engineering model, based on the physical reality, it is of such precision to ensure a safe flight. When we take the umbrella and the raincoat after watching the weather channel, we "trust" the weather forecast because it is based on models tested over appropriate space and time domain. When dealing with climate prediction, the models cannot be tested with non-existent data (Because they refer to the future). In practice, the virtuous circle forecast—verification of data—cannot be closed. The maximum we can do following the advise of Tim Palmer is to test the "fast physics" of the model that is the part of the model that coincide with weather forecast but we cannot test the "slow physics" (as advocated by Carl Wunsch) because the data are not available.

A possibility remains related to induction and for this specific aspect, we will refer to the work of Frame et al. (2007) Following the definition by Bertrand Russel: the induction is a process through which we can generalize our vision of the world based only on few observations. We already made some examples in the previous chapter, but it is worth to insist a little bit. We have seen that based on the induction, we can formulate general laws based only on observations made in a given moment. These laws can then be applied to the design of machines (like planes for example) or, if they were to exist, to predict the future climate. The problem of induction was also described by the English philosopher David Hume in his *Treatise of Human Nature* published in 1739 and from it we can extract the first idea about models. These should be used following Frame et al. (2007).

In other words, we have to use models, maps or other conceptual schemes to move from observations of regularities to our portrayals of causal connection. In order to develop those conceptual schemes, we need to: (i) imagine potential schemes and (ii) make decisions regarding their applicability or relevance. Developing scientific models is, in this sense, an imaginative exercise: we try to infer causal relationships, and then try to isolate, quantify and test these. Sometimes this is more difficult than other times. It tends to be much simpler when the systems under study are simple or well understood or time independent. It tends to be harder when isolation is elusive and where testing is difficult, as is the case in climate research

We have used the verb "infer" that will be central in all that we will use later. This process is sometimes simple and sometimes is very complex, as in the case of climate research, because in this case the relationships to isolate are elusive and even more complex is to test them. A typical example is the so-called "detection and attribution" problem where a specific meteorological event must be related to the climatic change.

In practice, the warming observed in the twentieth century has been interpreted not as a natural phenomena but rather has been attributed to the human activity. In this case, the radiative forcing due to the greenhouse gases and volcanic eruptions is known with some approximation (they are theoretical values anyway) and its effects can be evaluated even with a simple model. Nobody ensures that there would be other possible explanations and especially if the same method could be applied to the past or future climate. Just to follow Hume, there are several other reasons for the wrong explanation. A very important factor is the constancy of some of the conditions. For example, the climate sensitivity could be constant only for a short period of time, so that it would be different during the ice ages and could again change in the future with a completely different carbon cycle.

The constancy argument is relevant to a more general problem and that is the constancy of the natural condition. There is a famous example due to Bertrand Russell (2007) given in his book The problems of philosophy. Russell makes the example of a chicken fed daily by a farmer. The chicken using the induction thinks that this will go on forever until one day the farmer wrung his neck. The chicken makes the hypothesis of the uniformity of nature (the farmer) which does not include the final surprise. For the chicken such a catastrophic event corresponds for an earthling to evaluate the chances for the Earth to be hit by an asteroid. In this perspective, Leonard Smith of the London School of Economics speaks about a "rosy scenario" (Smith 2002) which respects essentially two constraints: (i) nothing happens that takes the model outside the range of its validity (that is there are not asteroids colliding with the Earth) and (ii) that all the significant feedback mechanisms are present. This is what is also expressed by Narcy Cartwright when in her book (Cartwright 2008) affirms that when laws are applied, they are valid, *ceteris paribus*. Translated for our case means that models are applied with constant conditions. We are not sure at all that this rosy scenario exists, because we have only an approximate knowledge of the interactions between the different climate variables, and especially considering the specific future inference, we do not know how the same variables will evolve in the future. Even excluding the extreme event of an asteroid impact, the oceanic circulation could change drastically for example with the increase of the carbon dioxide concentration in the waters. As Frame et al. affirms, the chicken did not have any chance to insert such a catastrophic event as the pull of his neck in his models. It could be extremely dangerous, as observed by someone, to infer something (the real world) with experiments based on an imperfect substitute of the reality (the models).

6.2 The Bayes Inference Applied to the Climate

In this chapter, we will deal with the application of the Bayes statistics to study the climate models. A very general introduction to this problem may be dated to the early 1990 s with a paper by the late David MacKay, a professor of engineering at Cambridge University and at that time brilliant Ph.D. candidate at Berkeley. Prof. Mac Kay is much more famous for his book *Sustainable energy, without hot air*. The hot air in this case refers to the heated debate which involves all the speeches on the future of energy. As a matter of fact in the presentation there is an unequivocal

This is a straight-talking book about the numbers. The aim is to guide the reader around the claptrap to actions that really make a difference and to policies that add up.

Good, David Mac Kay at that time has the same clear ideas about Bayes. In 2003, he published a heavy treatise (that was available before that on the net) with a complex title, *Information theory, inference, and learning algorithms*, Mac Kay (2003). The chapter that interests us most is 28 with the explicit title, *Model Comparison and Occam's Razor*. In this chapter, Fig. 28.4 (reproduced here as Fig. 6.1) could have been extracted with very low probability from an IPCC report while it has a strong resemblance to a climatic program proposed by Richard Goody and coworkers in a paper appeared on the June 2002 issue of *Bulletin of the American Meteorological Society*.

The process is very similar to the procedure "science" should follow in general that is, after getting the data, use a model to interpret the data. We are supposed to study a simple system. The so-called inference refers to the two boxes with the double edge. The first one finds the parameters that one uses in the model to maximize the compatibility with the data. This procedure can be followed with other methods than Bayes. On the other hand, where Bayes is good is in the second box where models are chosen based on their comparison with the data. This second process is much more complex because it must compare complex models that use more parameters (to explain a larger data set) with models that explain a more limited data set with more completeness. To better understand this part, we need the Bayes theory which is introduced in the Appendix A.

We suppose to have a number of model with M_i indicating the i-th. This model depends on a series of parameters w_1, w_2 w_3, etc., that made up the vector **w**. The results of this model can be found with a probability distribution $P(w \mid M_i)$ called the "prior". This model will make a prediction on the data D with a probability distribution $P(D \mid w, M_i)$. An example again taken from a paper by Stephen Leroy et al. (2015) will clarify everything.

Imagine then to have predicted the sea temperature at 13 °C. By changing the model parameters, we do not obtain always the same temperature but values which are distributed according the dashed curve shown in Fig. 6.2. The shape of the curve is the familiar bell curve that in mathematical terms is known as a Gaussian. The width of this curve is related to the parameter "sigma" which gives an idea of the dispersion in the temperature prediction and in this case, we have chosen 2.7. In

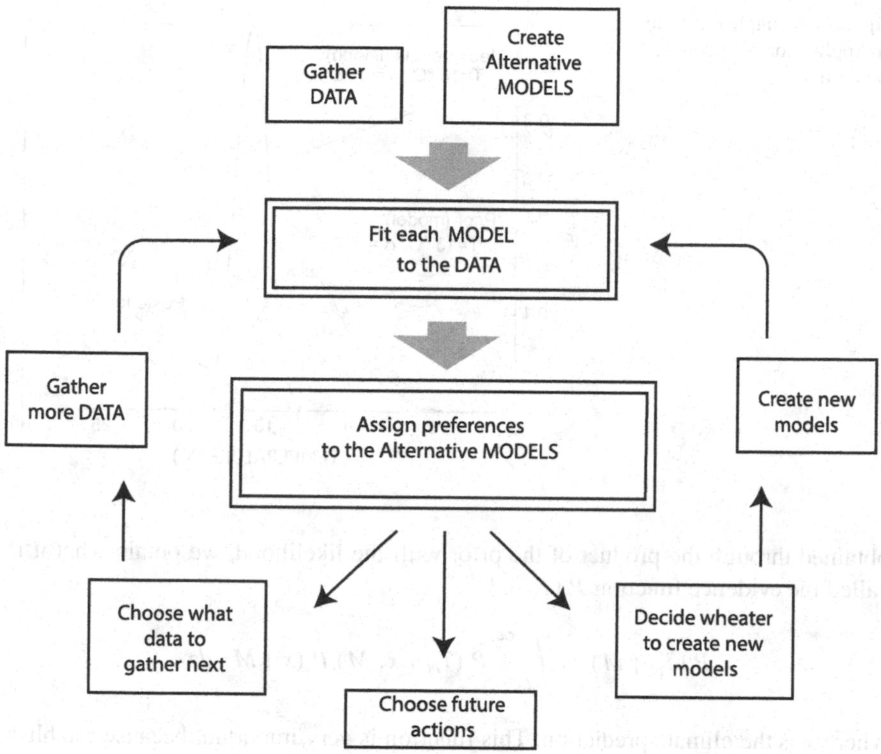

Fig. 6.1 The abstraction of the process in which data are collected and modeled. The double framed boxes denote the two steps which involves inference (Redrawn from Mac Kay 2003)

practice, this is equivalent to say that the model predicts a temperature of 13 °C with an uncertainty of 2.7 °C. Then we make a measurement and find an average value of 16 °C with a dispersion of 1 °C as shown by the Gaussian with the solid line. The dashed curve gives the probability to predict a given temperature in a given interval. The probability is at the maximum for the central temperature and decreases more and more when we move away from the central value. The same thing happens for the measurement: the most probable data is around 16 °C and in practice if we indicate with $P(T \mid M)$ the probability to find the value T predicted by the model M also means that the probability of a temperature between T and $T + \Delta T$ is just $P(T \mid M)\ \Delta$T. The same thing is true for the measurement so that the probability would be $P(T_m \mid T)\ \Delta$T to find a value in the interval ΔT with an average value T_m; when we apply the Bayes formula the posterior probability would be $P(T \mid T_m, M)$ proportional to $P(T \mid M)\ P(T_m \mid T)$. As seen in the figure this distribution is still a Gaussian with the maximum slightly shifted to 15.63 °C respect to 16 °C with a width of about 0.35 °C. All these results can be obtained with the classical statistics but with Bayes. We can make a step further and that is if we sum all the probabilities

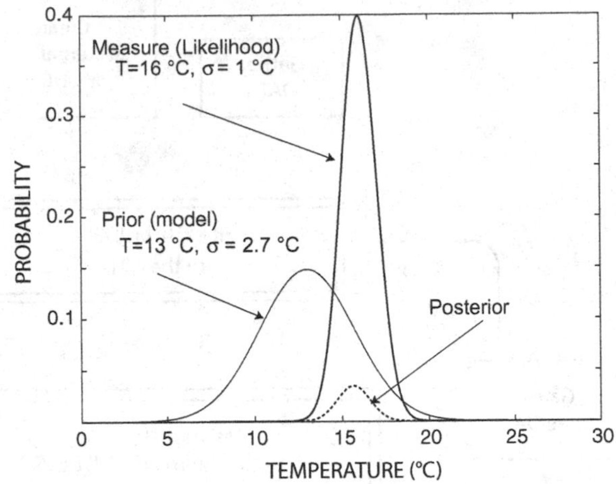

Fig. 6.2 A simple example of Application of Bayes theorem

obtained through the product of the prior with the likelihood, we obtain what it is called the evidence function $P\left(T_m \mid M\right)$.

$$P\left(T_m \mid M\right) = \int_{-\infty}^{+\infty} P\left(T_m \mid x, M\right) P\left(x \mid M\right) dx$$

where x is the climate prediction. This function is very important because establish the exact dimensions of the posterior probability that is the complete formula

$$P\left(T \mid T_m\right) = P\left(T_m \mid T\right) \times P\left(T \mid M\right) / P\left(T_m \mid M\right)$$

Which can be read as

$$Posteriorprobability = Likelihood \times Prior/Evidence$$

The formula can also be read in another way where $P\left(T \mid T_m\right)$ is the probability to find the temperature T given the measured value T_m. $P\left(T_m \mid T\right)$ is the probability that the data T_m is found given the distribution T, while $P\left(T \mid M\right)$ is the probability to find the temperature T as a result of model M. The evidence function in the present simple case is made of two terms. The first one depends on the precision that is from the half widths of both the distribution of the prediction and the data. The other term is the accuracy which takes into account the difference between the prediction and the measurement (13 °C respect to 16 °C) and from the sum of the square of the half width. Ordinarily, if the accuracy is a number greater than 1 either the model or the data are discarded. A simple application of the evidence function may be based on its evaluation for two models M_1 and M_2. The ratio of the two functions gives an indication on the goodness of one model with respect to the other if the ratio is greater than 1. In practice, the figure can be interpreted again in another way. If the prior

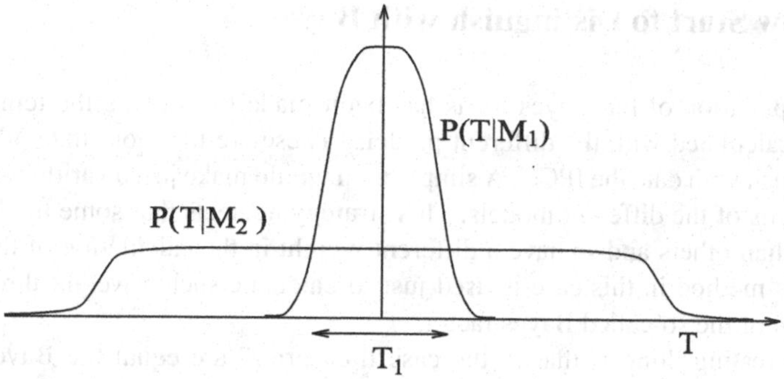

Fig. 6.3 The reason why Bayes theorem include the Ockham razor

Gaussian is very wide, this means that at the change of the parameters in the model the prediction involves a large range of temperatures and vice versa as shown in Fig. 6.3. As an example the model M_1 the prediction involves a limited temperature field while the model M_2 makes a prediction on a larger field of temperature. Model M_1 is necessarily simpler than M_2 because its prediction is much more limited. If the measurements were to confirm the prediction in the region of M_1, this would be also the most likely model. The reasoning can be put in a more precise form. We can relate the plausibility of model M_1 given the data $P\,(M_1 \mid T)$ and the prediction made by the model about the data $P\,(T \mid M_1)$ and the prior plausibility of M_1 $P(M_1)$. This gives the probability between M_1 and M_2

$$\frac{P\,(M_1 \mid T)}{P\,(M_2 \mid T)} = \frac{P(M_1)P\,(T \mid M_1)}{P(M_2)P\,(T \mid M_2)}$$

The first ratio on the right-hand side reflects our initial belief but the second ratio expresses how well the observed data were predicted by M_1 with respect to M_2. This formulation embodies the so-called Ockham razor. As shown in Fig. 6.3 a simpler model(M_1) will make a prediction on restricted range of T while the more complex model (M_2) will spread its predictive probability over a wider range. Based on the previous formula over the range T_1, the simpler model will be much better than the complex one.

The apparent conclusion seems to indicate that the best climatic model is the simpler one (for example, an energy balance model) but this cannot be a general criteria. As a matter of fact, if we want to make small-scale prediction (regional) such a model is not very useful. Then when we talk about Bayes application to climate modeling the reasoning must be little more sophisticated. However, this could be independent on the Occam razor argument because in a complex field like climate there could be many doubts on its use.

6.3 How Start to Distinguish with Bayes

A first application of the Bayes ideas has been made to evaluate the temperature averages calculated with the different models. These are the more than 40 models that refurnish with data the IPCC. A simple man would make just an arithmetic mean of the results of the different models. This strategy neglects that some model could be better than others and so have a different weight in the calculation of the mean. The Bayes method in this case is used just to calculate such a weight through the calculation of the so-called Bayes factor.

An interesting thing is that in the case the "prior" are equal the Bayes factor depends only on the likelihood functions. Suppose to have a model M_1 that should reproduce the set of data D. The probability of such happening $P\ (M_1 \mid D)$ is proportional to the prior $P\ (M_1)$ multiplied by the evidence function. In practice then the Bayes factor is the ratio of the likelihood functions. As we have seen, these depend on the distance between the calculated and observed values and the correlation existing between prediction and data. At this point, the Bayes factor is an indication of the model "goodness" with respect to data. But here we are in trouble because while it is easy to calculate everything for the model (averages, deviations, etc.) very often the data cover a period of time which is not sufficient for the model prediction. We may have mentioned the ECMWF program which covers about 50 years (ERA) and a similar program originated at the National Center for Atmospheric Research (NCAR).

The are several attempts to deal with this situation. For example, Seung-Ki Min (at the time of the work at the Ontario University) Min (2007) used 20 years of data to compute the Bayes factor for three different scenarios. The first is the control case and refers to a non-perturbed model; the second refers to a model perturbed only by the greenhouse gases and the last is perturbed both by the gases and the aerosols. The criteria adopted by Min is such that models with Bayes factor less than 1 have very low probability. The data are for the 2 m temperature and lower stratosphere temperature and cover the period between 1980 and 1999. It is obvious that the real interesting period is in 1990 and actually the control run shows very low Bayes factor, and this seems to exclude that variations in the last years of the 1990s could be due to natural causes.

Another application of the Bayesian statistics refer to the averages. The IPCC conclusions are based on a fixed "park" of model (a few dozen) that only apparently have the same characteristics. Usually in the presentation of data, averages are performed and this means to treat democratically the different models that is equivalent to give the same weight to all the models. This assumes that each models represents a truth.

The fact that for each model the Bayes factor can be evaluated gives the possibility to make weighted means for the different geographic regions on the Earth. The final results is that in the century that goes from 2000 to 2099 the situation looks reasonable for the first 40 years so that the temperature changes are distributed in a Gaussian fashion and the model groups around an average value. In the following decades, the variations group around two or three values giving a distribution that is known as multimodal. This fact raises some suspicion not so much on the method adopted for the means as on the quality of the models. It is like if the model predictions had more than a solution, that is, we would have something like a multiple equilibria solution. This kind of solution usually results from nonlinear system but the climate modelers are never worried with this aspect so that the reason for the multiple solution must be found somewhere else, for example, because there are too few a models for the application of the statistics.

The last application of the Bayes statistics takes us back to a method invented many years ago by the founding father of the stochastic models, Klaus Hasselmann. This is the fingerprint method which assumes that every climate change should leave its own fingerprint.

6.4 Fingerprinting the Climate Variations

Even if we admit that the internal mechanisms of the models are correct, the definitive proof of their predictive capacity would be the detection of a climate signal. This is not a simple problem even if we reduce it to a discrimination between signal and noise due to the natural variability. For example, how do we know that the warming observed in last century is due to the increase of the carbon dioxide and the other greenhouse gases?

A possible method is to compare the distributions of the calculated temperature changes with those observed up to the present epoch in two cases: the heating due to the increase of solar radiation and the same heating due to the changes of the greenhouse gas concentration. As we can see from Fig. 6.4 the distribution in the temperature changes in the two cases is completely different. If the temperature increase is due to the carbon dioxide, then to an increase in the tropospheric temperature corresponds a cooling of the stratosphere. In case the increase in temperature is due to the solar radiation we observe a warming everywhere. The visual fingerprint in this case is quite clear and would be possible to distinguish between the two cases. In general, however at the present time, the data are not clear enough and especially for the natural fluctuations it is such to require more sophisticated methods. One of these methods is the so-called *optimal detection*. A simple example of this method is due to Gerry North of the Texas A and M University. Suppose to estimate the temperature of a pond taking measurements with two different thermometers. The value of the temperature will be the weighted average of the two measures that will have an average value and a standard deviation. The weights of the two measurements will be obtained by minimizing the standard deviation of the weighted mean. We can then find that the optimal measurement of the temperature is a simple function of the single temperatures of the two thermometers weighted by the respective standard deviations.

One of the problem of detecting a climate signal is that we know (or pretend to know) the signal we expect, but we do not know its entity. For example, we know that

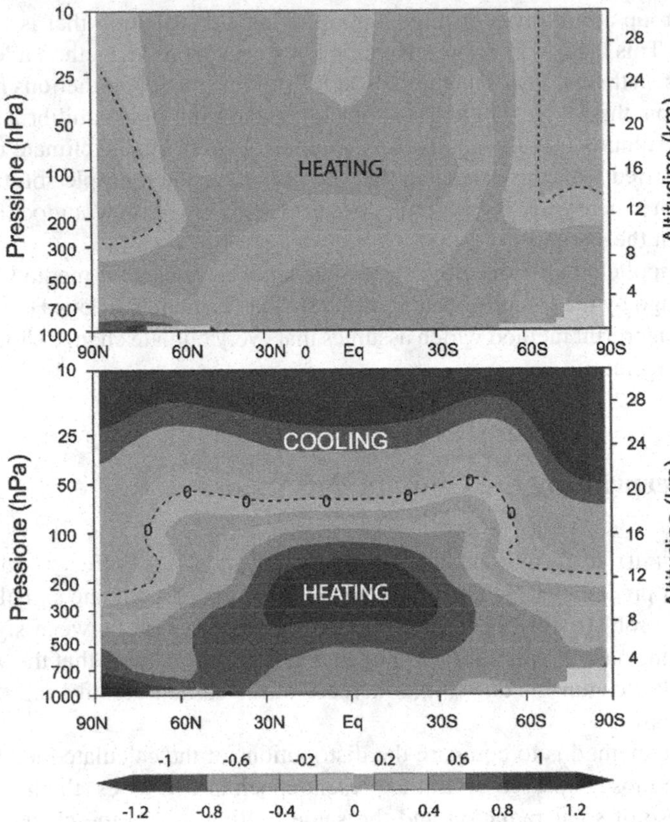

Fig. 6.4 The temperature variations due to an increase in solar radiation (*top*) compared to changes due to doubling of carbon dioxide (*bottom*). The dashed line separates the cooling and the heating regions

the average temperature will increase but we do not know what the increase would be exactly in a particular region. This of course is a fundamental problem because of the detection of the signal needs that it has to be discriminated from the noise. We try to make a simple example extracted from one of the IPCC report. Imagine then that the surface temperature (one of the many climatic variables) depends from two different processes, the amount of greenhouse gases (the forcing) d_1, and the changes in the solar radiation d_2. The "noise" that is the natural variability associated by these two process can be represented on the plane d_1,d_2 by an ellipse (n). All the possible combination of the variables due to noise are comprised in the elliptic area as shown in Fig. 6.5. The signal can be detected only if it falls outside this ellipse. We can then trace two ellipses that represents the regions where we expect the signal from the two mechanisms. If the signal would be of type (a) then it could acceptable within mechanisms (1) while a signal like (b) could be acceptable both for mechanism (1) and (2). On the other hand (c) is not acceptable because it falls outside any area

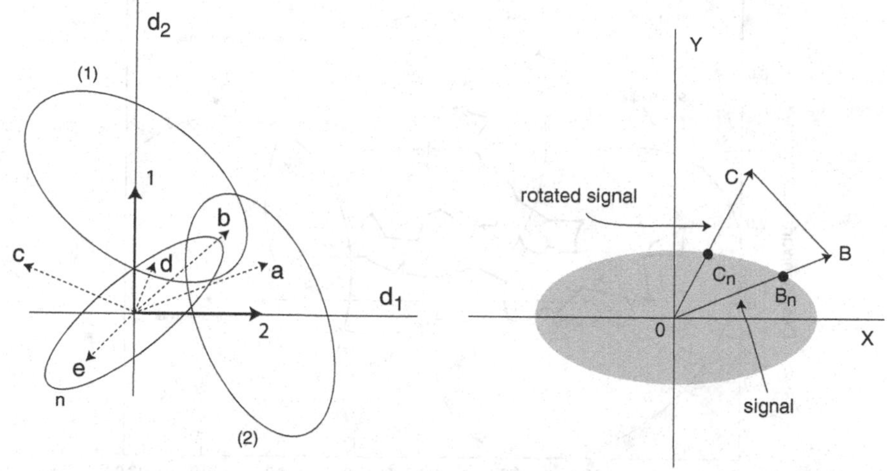

Fig. 6.5 Optimal detection. The ellipses on the left refer to the noise (n) and two distinct mechanisms (1) and (2). The signal are a, b. The gray ellipse on the right represents the noise area. A signal B has the ratio with noise B/B_n while if rotated has a ratio C/C_n

delimitated by the two mechanisms while both (d) and (e) are within the natural variability. As of today this remain largely an academic speculation because it is based mostly on the model results and not on real data. The key of the method is just to reveal a signal in the best of conditions. If we now look at the figure on the right we see that a signal which remains outside the noise ellipse when rotated assumes the best signal/noise ratio. This is actually the optimal detection, that is to find the conditions that maximizes the signal-to-noise ratio. An application of the optimal detection is shown in Fig. 6.6 where the ellipses of the previous figure represent the mechanisms of the changes of the solar constant, of the greenhouses gases and greenhouse gases and aerosols. A preliminary analysis shows that the signal is compatible mainly with the mechanism aerosols and greenhouse gases. In the Appendix, we will discuss a little further this problem.

Someone supported the idea that fingerprinting is the methodology needed to study process modeling rather than climate modeling. Both objectives need good climate data but the optimum data will differ. Thus the design of a climate observing system must be studied in the light of this requirement. To get the climate community aboard with new attitudes may require emphasis on finger printing when planning observing systems.

To understand what Bayes has to do with the fingerprinting we must recur to a more general argument. Imagine to have a map with the changes in temperatures from observations measured in the last 50 years with the map referring for example to Europe. The observations refer to a grid and then to a matrix of values. This matrix can be related to a variation of the average temperature distributed on the same grid according to a matrix g,

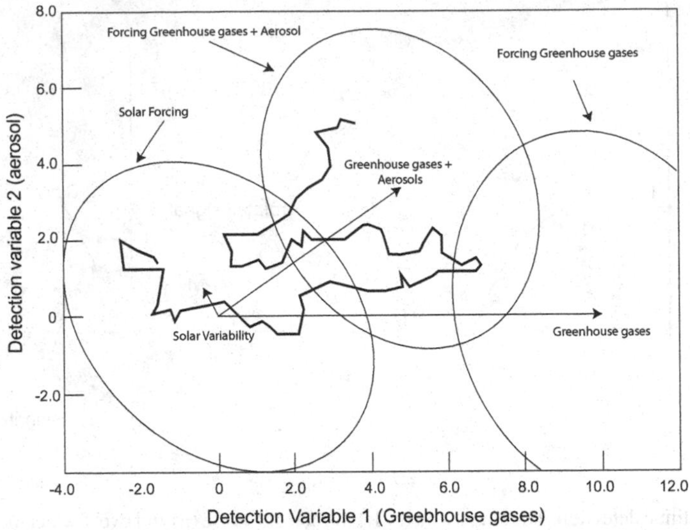

Fig. 6.6 The application of the method of the optimal detection to the temperature record (*broken line*) from 1946 to 1995. (redrawn from Hasselmann 1998)

$$Observations = g \times a + noise$$

In practice, the models are necessary to determine both the noise and the distribution matrix g of the amplitude a. The Bayes method in this case is used for a better determination of a. An a priori estimation can be obtained with classical methods (for example what statisticians call best fit). Once a preliminary value is obtained for a it is possible to improve it with the Bayes method. This is well illustrated in Fig. 6.7 where it is evident how the estimation of a is improved to the point it coincides with the climatic changes.

This study made by three known statisticians Mark Berliner of the Ohio State University, Richard Levine of the University of San Diego and Dennis Shea of NCAR, is also interesting because it can be interpreted as a further proof that the global warming may be attributed to human activities. The forcing is represented as in Fig. 6.7 with the prior made by two components. The first one is centered around "zero" which represents the natural forcing (in the absence of greenhouse gases) with a width related to the natural variability. The second component is due to the forcing of the greenhouse gases with a width related to the variability of the anthropogenic forcing.

This prior is practically constant in the four decades chosen for the simulation that correspond to 1961 to 1998, and 1970 to 1998, from 1980 to 1998, and from 1988 to 1998. The likelihood function presents a progressive increase of the temperature "shift" that becomes higher as we approach the last decade. The likelihood function is based on the simulation obtained with the GCMs. What is to note in the posterior

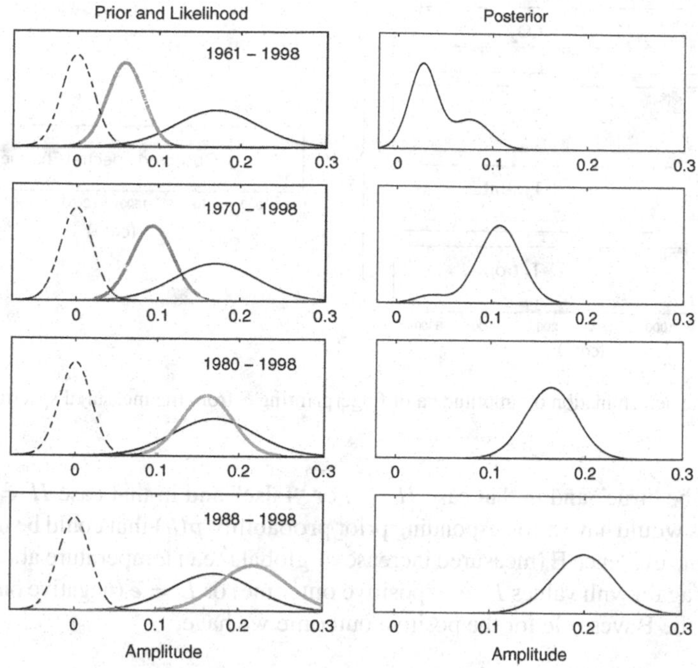

Fig. 6.7 The result of the simulation by Mark Berliner and his coworkers. The column on the left shows the "natural forcing" (dashed line) and the anthropic one (thin solid line). The likelihood function is the thick gray line while the posterior probability is on the right (redrawn from Berliner et al. 2000)

probability is the progression of the warming which is absent in the first two periods which reflects the natural variability. The Bayes statistics is then a very useful instrument for the climate studies and there is a last application which can help to decide which data are better for the climate prediction. This topic however needs a chapter by itself.

A similar approach was developed by Jim Anderson group at Harvard (Leroy et al. 2008; Huang et al. 2010) and applied to the spectral signature that we mentioned in the previous chapter. As shown in Fig. 6.8 in this case it is presumed to measure the spectral changes resulting from an increase in carbon dioxide and this constitutes the vector **y**. The fingerprinting **S** on the other hand are the single physical contribution to the signal and the relation between these quantities is again

$$\mathbf{y} = \mathbf{Sa} + \mathbf{r}$$

Figure 6.8 gives a very schematic summary of the process.

Klaus Hasselmann in 1998 paper (Hasselmann 1998) has tried to combine the fingerprint method with Bayes but he got so and so results. He observes first of all that given the hypothesis $H = \overline{h}$ (there exists an anthropogenic GHG forcing)

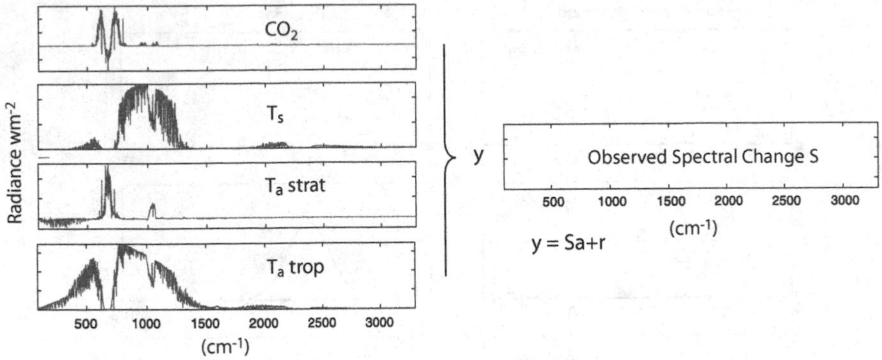

Fig. 6.8 The determination of amplitude **a** of fingerprinting **S** from the measured spectra **y**

this could be "true" and in that case $H = h$ or "false" and in that case $H = \overline{h}$. This hypothesis would have a corresponding prior probability $p(h)$ that could be modified based on the evidence E (measured increase of global mean temperature above some level) and again with values $E = e$ (positive outcome) or $E = \overline{e}$ (negative outcome). Applying the Bayes rule for the positive outcome we have,

$$p\,(h \mid e) \times p(e) = p\,(e \mid h) \times p(h)$$

Now we introduce a variable c for "credibility" as the posterior probability $p\,(h \mid e)$ corresponding to the prior $c_0 = p(h)$ while $l = p\,(e \mid h)$ is the likelihood function. It is possible to express (see Appendix) the credibility c as a function of the ratio $\widehat{l}/l$ where $\widehat{l}$ is the complementary conditional likelihood $p\left(e \mid \overline{h}\right)$. It is found that in the classical approach the credibility decreases linearly with the ratio $\widehat{l}/l$ while in the Bayesian case the credibility strongly depends on the prior probability reaching c_0 for values of $\widehat{l}/l = 1$.

Finally there is probably an intuitive explanation for the detection ellipses we drew in several occasions and the distribution of the climate variables of forcing functions. If we assume a model ensemble subjected to a single forcing (i.e., increase of carbon dioxide) the results are distributed (maybe) in a Gaussian way around the mean. The same ensemble when forced with aerosol may show a similar distribution but with different parameters. The combinations of the two distributions for an assumed probabilities will give an ellipse whose axis will be oriented in a number of ways with respect to the axis: carbon dioxide-aerosol. This orientation will depends on the degree of correlation of the two variables.

6.5 Prioritizing the Data

At this point, we can make some consideration about the possibility that all this movement looks more like a huge academic circus. As a matter of fact in the Appendix we show that the Bayes statistics has many useful application for everyday life as it is shown for example in the medical field. In the realm of climate studies, the applications are numerous although the data are scarce. In what follows, we will see that the Bayes statistics can suggest which data are better. However the government (but mainly the practitioners of the field) must convince themselves that the first priority remain the data.

Not all climate measurements are equally useful for improving climate predictions, and the ability to choose between data types on the basis of value is useful both for allocating observing resources and for the efficient use of computer resources. But before discussing this topic, we need to take note of the fact that both predictions and observations are uncertain to some degree. When we compare them we are comparing two uncertain quantities and our conclusions will also be uncertain. The appropriate mathematical framework for handling such quantities is the science of statistics, a discipline with its own assumptions and methods, and there is increasing belief that, for assimilating climate data, the appropriate system of statistical inference is that of Bayes. Bayesian inference treats relationships between probability density functions (pdfs), quantities which express the probability of occurrence of a quantity in terms of the value of that quantity (see Fig. 5.1 for examples of pdfs of climate trends). With only 38 results in Fig. 5.1, it is only possible to describe the simplest possible pdf (a Gaussian pdf) which has the same mean and the same standard deviation (the spread of the pdf) as the data. To illustrate how Bayes work, we assume that the pdf of the Fig. 5.1 is a Gaussian as that represented in Fig. 6.9 and indicated as "prior", i.e., the information that exists prior to the introduction of new evidence. Now suppose that new evidence becomes available (the horizontal arrow in Fig. 6.9), this adds to our knowledge about the climate, and requires that the prior be modified to the posterior. The maximum of the posterior must be closer to the truth than the prior, because it is based on more evidence, and it should be more certain (smaller spread). Bayesian inference put these statements on a quantitative basis.

Now the problem that Huang et al. (2011) have formulated is to use Bayes method to estimate which data are the best to detect global warming.

The statistical problem is that of an ensemble of pair x, y generated by an ensemble of climate models E where the x variable refers to a quantity that can be observed and y to a quantity that can be predicted. Then the prior is assumed for simplicity to be a gaussian $P(x, y \mid E)$ while the likelihood function using data d depends only on x, $P(d \mid x)$. We apply the Bayes theorem to find the posterior $P(x, y \mid d, E)$

$$P(x, y \mid d, E) = P(d \mid x)\, P(x, y \mid E) / P(d \mid E)$$

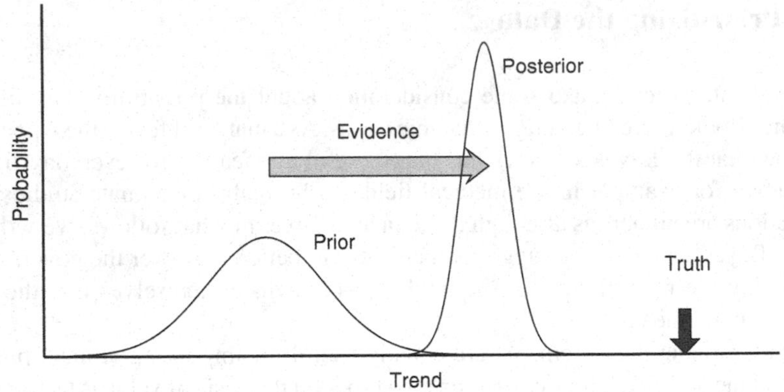

Fig. 6.9 A schematic illustration of Bayes inference. Goody and Visconti (2013)

$P(x, y \mid d, E)$ is called the posterior and represents the ensemble prediction *x,y* modified by the data. The likelihood function $P(d \mid x)$ is the likelihood of the data given the a predicted observation. The prior $P(x, y \mid E)$ is probability distribution of the ensemble before the data are considered. Now for simplicity, we drop the ensemble symbol and assume that the probability of prediction of *y* is true $P(y \mid d)$ is given by

$$P(y \mid d) = \int P(x, y \mid d)\, dx \propto \int P(x, y) P(d \mid x)\, dx$$

The work of Huang et al. implements the outlined method by assuming that the distribution of the variables *x,y,d* are Gaussian with averages μ and standard deviations σ. Because the data were not available, it was assumed that one of 21 members of the ensemble was the "truth" that is simulated the data. This model was the one from National Center for Atmospheric Research (NCAR) and the data to be predicted was the 50 year surface temperature trend that for this model was 0.019 K/year. The prediction *y* was the 50 years trend using the data *d* and the simulations *x* every 10 years. Figure 6.10 shows the results of a 50-year of globally averaged surface temperatures trend after adding information about the climate of the first 10 years, the first 20 years, etc. The maxima of the pdfs are shown by the points and arrowheads and the standard deviations by the vertical lines. The mean of the prior is represented by the empty square point; the filled square and triangle points and lines are the posteriors. Two types of data are shown: squares are for globally averaged surface temperatures;triangles are for certain radiances, measured from a satellite. The figure demonstrates the expected improvement of accuracy and reduction of uncertainty as new evidence is added, and shows, surprisingly, that adding space radiances may be slightly more useful than adding surface temperature data, even when the latter is the predicted quantity. These calculations can be extended to include many data types.

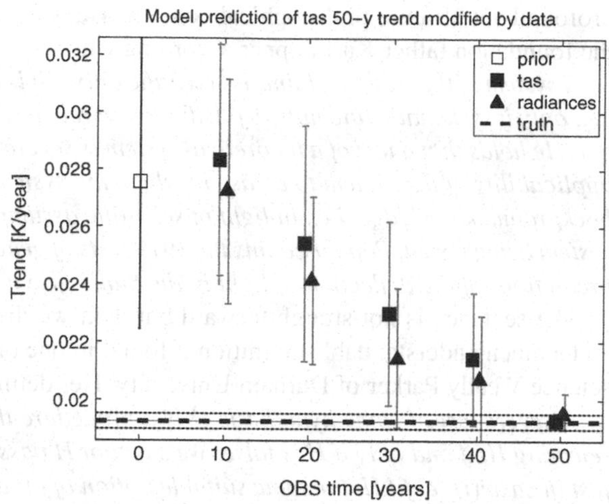

Fig. 6.10 Evolution of a prediction. Model predictions of surface temperature trends over 50 years are modified by data at 10, 20, 30, 40, and 50 years. Calculations for two data types are shown: surface temperature trends (tas), and trends in all satellite radiances taken together (radiances) (Huang et al. 2011)

Data types that stand out as being more useful than others for improving predictions of surface temperatures are to be preferred.

In the Huang et al. work different data types are used and a quantitative criteria is given for the improvement in accuracy, related to the average values of the prediction and the posterior, and the precision related to the difference of standard deviation. It is shown that the trend in the temperature and the in situ data give improvement of 0.21 and 0.35 respectively why the radiance observations and the all satellite data give improvement of 0.44 and 0.53 after 20 years of observations. It looks like CLARREO was the right tool to solve scientifically the global warming conundrum.

More recently Stephen Leroy (Leroy 2015) and coworkers have updated the work of Huang using the larger set of data provided by CMIP5 (the fifth exercise of the model intercomparsion project). The purpose was the select the best data set that could improve the prediction of future climate using the correlation between the "hindcast" for the period 1970–2005 with the prediction between 2090–2100. The variables were chosen in such way that could be measured and they were 200 hPa surface (roughly 12 km) that can be measured with the GPS, and the different type of radiation that can be measured from space. The result of this study were quite inconclusive because the very poor correlation between the climate prediction and the hindcast indicating one again the performance of models are quite poor.

6.6 Are There Alternatives to Bayes?

In 2011, the magazine *EOS, Transactions*, of the American Geophysical Union published a short note by Joel Katsav in the *Forum* section (Katsav, 2011). Katsav is at the Technology University at Eindhoven and with the "severe tests" was hiding a profound yet obscure philosophical concept. He defines the severe test referring to the foundation father Karl Popper according who

A system of theoretical claims is scientific only if it is methodologically falsifiable, i.e., only if systematic attempts to falsify or severely test the system are being carried out. He holds that a test of a theoretical system is severe if and only if it is a test of the applicability of the system to a case in which the system's failure is likely in light of background knowledge, i.e., in light of scientific assumptions other than those of the system being tested. Popper counts the 1919 tests of general relativity's then, unlikely predictions of the deflection of light in the Sun's gravitational field, as severe.

The sentence is not straightforward but then we discovered that it was not new and a much understandable definition if found in one of the paper by philosopher of science Wendy Parker of Durham University. Her definition reads (Parker 2008).

A severe test of some hypothesis H is a procedure that has a high probability of rejecting H, if and only if H is false. We say that H pass a severe test with results 'e' just in case:(i) 'e' fit H, for some suitable notion of fit; and (ii) it is very unlikely that the test procedure would produce 'e' that fit so well with H, if H is false. if H does pass a seasevere test with results 'e', then 'e' are 'good evidence for'- or 'a good indication' of—H. Put more informally: we have good evidence for H just in case a procedure that almost surely would have indicated H to be in error, were actually erroneous, nevertheless does not indicate that H is in error.

This definition is a little more clear but still there is no indication of what is intended for "severity". The bible of the severity approach is a 1996 book by Deborah Mayo professor of philosophy at Virginia Tech which presents plenty of discussions and few rather vague examples. A subsequent paper in 2010 clarify much better the idea starting from a detailed definition of the severity principle but does not give again examples that clarify what severity means. The major objection to the Bayes approach is again that made by Katzav in his EOS paper. He claims that the evaluation of the posterior probability $P\,(F \mid data)$ of a parameter F based on data, the same data do not *provide seasevere tests of estimate of F*. And again *this approach does take into account the extend to which model simulation agree with the data—something that is captured by* $P\,(data \mid F)$ (i.e., the likelihood function) *but we have seen that degree of agreement with data is not itself a measure of test severity*. Katzav also claims that the severity test clash with another simple Bayes rule. In this case if one has a very low prior $P(F)$ this results in a even lower posterior. On a seasevere testing approach the confidence in the value of F should increase with the test severity and so even low initial probability should increase after the test. Katsav also claims that the normalization factor $P(data)$ used in the Bayes formula $P\,(F \mid data) = P\,(data \mid F)\,P(F)/P(data)$ is the same for all the F and it cannot be an indicator with which data test different estimate of F.

This is plenty of beautiful theory but then we look at its application to climate modeling. This should be the subject of a 2013 paper Katsav (2013), which contrary to the expectations gives a very qualitative application of the severity principle. The hypothesis where to start (called OUR FAULT) claims that the *increases in anthropogenic greenhouse gas concentrations caused most of the post-1950 global warming*. and this hypothesis passes the severity test after a simple visual inspection of a couple of figures extracted from IPCC fourth report and the conclusion is the following:

From Mayo's perspective, OUR FAULT has passed a severe test with observed post-1950 GMST trends if the simulations of these trends that are consistent with OUR FAULT fit the trends and:

(1) it is very unlikely that, had OUR FAULT been false, the simulations that are consistent with OUR FAULT would have fit the trends as well as they do.

(2) will be established on the basis of the observations being appealed to only if they are used to rule out the alternatives to OUR FAULT. In the present context, this means that, for (1) to be established, the IPCC-AR4 simulations of post-1950 GMST trends that are not consistent with OUR FAULT—e.g., those in which only natural variability is assumed to be operative—must be used to show that, for each alternative, it is very unlikely that OUR FAULT would be as successful at capturing the trends if the alternative were true.

At the least it is disappointing that a severity test is performed in such a qualitative way. We have illustrated so far how articulated is the effort to validate the models by comparing their results with data. Yet there is another aspect that shines through in the second statement and that is the possibility of alternative hypotheses to the global warming. But soon Katsav explains that this problem cannot be solved with the IPCC-AR4 because alternative hypotheses do not enter in the IPCC simulation.

This is another cue ball of the philosophers that Bayes could eliminate alternative hypothesis to an accepted theory and in this respect is very instructive to read the amusing book by John Earman, (Earman 1996). In Chap. 7 he discusses the alternative hypotheses to the General Theory of Relativity (GTR) that was already discussed in terms of severity test. This must be mentioned as a useful example because of the two classical tests for GTR (the motion of Mercury perihelium and the bending of light by massive stars) the first is recognized as the most severe. This is quite a problem for Bayesian supporters because the Mercury motion was known before the GTR was formulated and someone suggest that Einstein may have used the data to "adjust" the theory. As for the discussion in Chap. 7 there is an interesting discussion how Bayes could help to decide which theory was right and discard alternative theories. When we transport such concepts to the climate debate we easily find that there are no alternative theories, because there is not a theory of climate with equations and so on. Mostly the present debate about climate focuses on the human responsibilities for the observed warming and these are strongly related to the GCM results.

The same group of philosophers also deal with the problem of climate fingerprinting. In particular this aspect is discussed by Wendy Parker. She affirms that a fingerprint is said to be detected at a confidence level of 95% if the analysis indicate that no more than a 5% chance that the amplitude of the fingerprint in observations would be produced by internal variability alone. Parker also notice that there is difficulty in accepting the fingerprint concept because in the particular case of the global warming, there is no data set referring to a situation in which greenhouse gases are absent. If this was the case one could compare easily the modeling with and without GHG. Also Parker notices that the simulations for fingerprinting, assume that the system is additive that is the different forcing are supposed to add up which is true as long as the system behaves linearly. Finally Parker affirms that fingerprint studies are at a very low confidence level (10 or 5%) so that the effects of variability dominate. With these several limitations in mind Parker concludes (Parker 2010).

Climate change fingerprint studies rely on some questionable assumptions, including the additivity assumption and the adequacy assumption, and are subject to a number of uncertainties. Nevertheless, results obtained so far are robust in their support for the conclusion that most of the observed global warming of the late twentieth century was caused by increased greenhouse gas emissions. Moreover, these formal detection and attribution studies focusing on near-surface temperature are only one source of evidence; observed changes in other parts of the also favor the conclusion that increased greenhouse gas emissions are the dominant cause of recent global warming.

In one of its very enjoyable books Steve Weinberg (Weinberg 1993) at chapter seven (*Against philosophy*) affirms ... *it is just that knowledge of philosophy does not seem to be of use to physicist always with the exception that the work of some philosophers help us to avoid errors of other philosophers.* Here we had some example which reaffirms the middle of the road political position of philosophers of climate.

Bayes Statistics and the Climate

A climate scientist and a humanist at the end of each chapter will discuss what we intended to say and what they have understood. The humanist in this case is more familiar with the many philosophical implications.

H: It remains very comforting that again there some kind of philosophical approach although the beginning recall a discussion of many years ago provoked by the idea of "science as a social construct". Those sentences on the intrisic rightness of the physics theories that are used to design aircraft seem to evoke criticism that fortunately does not have consequences.

C: I do not want to repeat what we already discussed but there is a phrase by Bertrand Russell (mentioned in this chapter) which say that "Science is what you more or less know and philosophy is what you do not know", In this chapter we give some detail on what could be the best methods to test climate predictions or at least to improve them.

H: It looks to me that a topic of particular importance is the uncertainty associated not only to the numerical models but also to the basic properties of the climate system.

C: Actually the present prediction in my opinion are a kind of dreamlike exercise because not only there are the model uncertainties but also those related to the economic future of the globe that has an obvious interaction with the political world. That is a prediction with a "physical" character is based on short term enormous uncertainties, never mind on a time horizon of 50 or 100 years. The economic collapse of the 2008 was not predicted so that is nonsense to imagine what will be the situation in 50 years. The uncertainties related to models are enormous as not simply those related to the use of the model but especially those related to non linear phenomena or other unknown phenomenology. An example in this direction are the models used to evaluate the ozone depletion. Toward the end of the 70's theories on the ozone depletion due to the presence of chlorinated compounds were based on homogeneous chemistry catalitic cycles, neglecting completely the effects of particulate. The discovery of the "ozone hole" was a complete surprise to the point that they insisted on the same chemistry for its explanation. The difference was the introduction of the heterogeneous chemistry suggested in 1986 by Solomon, Garcia, Rowland and Wuebbles. This proposal, confirmed by the priceless measurement by Jim Anderson (never the Nobel prize was more unfair) introduced a non linear chemistry that explained effects never predicted by any model. Today the same thing could happen with the climate, a catastrophic acceleration of the warming or the opposite. This without mentioning the alteration to the carbon cycle (methane or carbon dioxide). As Carl Wunsch repeats the ocean is the great unknown and could fill the role suggested by Shakespeare "There are more things in heaven and earth, Horatio, Than are dreamt of in your philosophy".
H: This strengthen the idea that science must be separated from the politics, because they have different times and certainties. So the politics must follow the road of prudence and low consumption while science must follow Francis Bacon advise, who recommended the road of patient research. The road that recommend to ask questions rather than proclaim answers, to collect proofs rather than hurries to judge, to listen to the voice of nature rather than that of the ancient sages. The program outlined with the Bayes statistics seems to respond to these recommendations.
C: There are a few problems to apply Bayes statistics to climate. If we examine a little the formula we notice that the "prior" can be assimilated to the model prediction if we refer to a particular parameter like the temperature. The likelihood function is "hard" because it implies to have an accurate statistics on the distribution of the variable. Finally the evidence function is the more important because it establishes the normalization, that is the absolute value of the posterior probability. This function is partly based on the model precision that is how much the prediction differs from the prior and partly on the accuracy which takes into account of the difference between measurement and prediction. models should be the better, the better the precision and the better the accuracy.

H: Some of these data already exist for example for the surface temperature so that you should already have some information on the quality of the models.
C: As a matter of fact a few model "ensemble" have been tested for the last 40 years of the past century but in a sense the results eat their tail. The likelihood function is produced with the help from models so that the results are rather qualitative. Some research shows that the last twenty years have temperature records that are not consistent with the natural variability. Bayes has been used also to evaluate the average of the prediction made by the models with another apparent negative results for the models. While in the initial period of the simulation the statistics of the averages are distributed according to a gaussian in period near the end of the simulation averages are distributed on more than a gaussian (multimodal distribution). This may point out to a serious problem for the models.
H: The fingerprinting method is only apparently a technical question while it looks more like a strategy that remain rather qualitative.
C: The fingerprinting method has been introduced in the '80 by Klaus Hasselmann that is not qualitative at all. This technique it can be shown that contains the Bayes approach and so may help to decide which kind of data are more suitable to detect a climate signal.
H: Again in the final section philosophers are back in the game but I am sorry to say with rather unfit statements.
C: When I first read the Katzav note on EOS I thought that this was a hard request made to the modeling community. But then I read a few papers by the same group and I could not find an acceptable definition for "severe test" and this discouraged me. It looks like the usual game by this people that stuck to some defintions that in this case are quite useless.

References

Berliner, L. M., Levine, R. A. & Shea, D. J. (2000). Bayesian Climate change assessment. *Journal of Climate*, *13*, 3805–3820.

Cartwright, N. (2008). *The dappled world: a study of the boundaries of science*. Cambridge: Cambridge University Press.

Earman, J. (1996). *Bayes or bust? A critical examination of Bayesian confirmation theory*. Cambridge: MIT Press.

Frame, D. J., Faull, N. E., Joshi, M. N., & Allen, M.R. (2007). Probabilistic climate forecasts and inductive problems. *Philosophical Transactions of the Royal Society A*, *365*, 1971–1992 by permission of the Royal Society.

Goody, R., Anderson, J., Karl, T., Miller, R. B., North, G., Simpson, J., et al. (2002). Why monitor climate. *Bulletin of the American Meteorological Society*, *83*, 873–78.

Goody, R., Visconti G. (2013) Climate prediction: an evidence-based perspective. *Rendiconti Lincei* doi:10.1007/s12210-013-0228-2. Reproduced with permission of Springer.

Hasselmann, K. (1998). Conventional and Bayesian approach to climate-change detection and attribution. *Quarterly Journal of the Royal Meteorological Society*, *124*, 2541–65.

Huang, Y., Leroy, S., & Goody, R. M. (2011). Discriminating between climate observations in terms of their ability to improve an ensemble of climate projections. *Proceedings of the National Academy of Sciences, 108*, 10405–10409.

Huang, Y., Leroy, S. S., & Anderson, J. (2010). Determining longwave forcing and feedback using infrared spectra and GNSS radio occultation. *Journal of Climate, 23*, 6027–6035.

Hume, D. (2003). *A treatise of human nature*. Dover.

Katsav (2013). Severe test in of climate change hypothesis. *Studies in History and Philosophy of Modern Physics, 44*, 433–41. Reprinted with permission from Elsevier.

Katsav (2011). Should we assess climate model predictions in light of severe tests? *EOS, 92*, 195–196. American Geophysical Union. Used with permission.

Leroy, S., Redaelli, G., & Grassi, B. (2015). Prioritizing data for improving the multidecadal predictive capability of atmospheric models. *Journal of Climate, 28*, 5077–5090.

Leroy, S. S., Anderson, J., Dykema, J., & Goody, R. (2008). Testing climate models using thermal infrared spectra. *Journal of Climate, 21*, 1863–75.

Leroy, S. S. (1998). Detecting climate signal: some Bayesian aspect. *Journal of Climate, 11*, 640–51.

Mac Kay, D. J. (2003). *Information theory, Inference, and learing algorithms*. Cambridge: Cambridge University Press.

Min, S. K., Simonis, D., & Hense, A. (2007). Probabilistic climate change predictions applying Bayesian model averaging. *Philosophical Transactions of the Royal Society A, 365*, 2103–2116.

Parker, W. S. (2008). Computer simulation through an error-statistical lens. *Synthese, 163*, 371–84. With permission of Springer.

Parker, W. S. (2010). Comparative process tracing and climate change fingerprints. *Philosophy of Science, 77*, 1083–94. Permission by University of Chicago Press.

Russell, B. (2007). *The problems of philosophy*. Cosimo Classics.

Smith, L. A. (2002). What might we learn from climate forecasts? *Proceedings of the National Academy of Sciences, 99*, 2487–92.

Weinberg, B. (1993). *Dreams of a final theory: The search for the fundamental laws of nature*. London: Vintage Books.

Chapter 7
Statistics and Climate

7.1 Introduction

Keeping in mind the famous definition of statistics (referred by Maynard Smith 1986) as that branch of mathematics which enables a man to do 20 experiments a year and publish one false result in *Nature*, nevertheless statistics has a growing influence in the climate science.

Climate science (if it exists) is contributed by a number of established sciences like fluid dynamics, thermodynamics, radiative transfer, chemistry, biology and so on. Until a few years ago statistics was not in the list but now thanks to a rather bully and qualified intervention of UK statistician, statistics must be included in the group of sciences that may give a decisive contribution to the climate studies. Crucial to this development was a resident course of the Isaac Newton Institute for mathematical sciences held from August to December 2010 with the title *Mathematical and Statistical Approaches to Climate Modeling and Prediction* whose proceeding were published on a special issue of *Envirometrics*. In a most recent paper on the new journal *Annual Review of Statistics and its Application* two of the authors of the revolution, Jonathan Rougier and Michael Goldstein explicitly affirms (Rougier and Goldstein 2014)

Now a full uncertainty assessment for climate policy appears doubly daunting. However, that is the wrong way to look at the problem. The really daunting task is to make and successfully implement a climate policy without doing a careful uncertainty analysis. Although these sources of uncertainty are challenging to assess, they are no more challenging than other parts of climate modeling (e.g., doing the basic science, formulating mathematical models, constructing simulators that run efficiently on supercomputers, and designing satellite missions to collect observations). The difference is that the climate modeling challenges are addressed by well-established communities. If climate policy is a genuine concern, then scientifically leading countries such as the United Kingdom need to develop a similar community of climate statisticians, working alongside the other communities and funded at a sufficient level, with the same access to computing facilities.

G. Visconti, *Problems, Philosophy and Politics of Climate Science*,
Springer Climate, https://doi.org/10.1007/978-3-319-65669-4_7

It looks more like a call to arms or rather the creation of a new community which has all the rights to say something decisive about the climate prediction. The aim of the statistician contribution to climate prediction is the reduction of the uncertainties or their quantification, topic quite unknown to the mainstream modelers. On the same issue of the Annual Review Peter Guttorp (2014) published a paper more focused on the data treatment but with the same spirit.

From a statistical point of view, it is appropriate to view the climate as the distribution (changing over time) of climate variables. These include, but as the quotation from Lorentz points out above, are not limited to, weather variables such as temperature and precipitation. The view of the climate as a distribution (and the weather as a random draw from this distribution) allows a statistician to utilize a substantial body of methodology and also indicates some directions of theoretical investigation (of empirical processes of multivariate nonstationary and temporally dependent observations). I take this point of view throughout this review. It is not one with which my climatology colleagues necessarily agree, as they tend to be empiricists and use definitions such as "climate is . . . the statistical description in terms of the mean and variability of relevant quantities over a period of time" (Solomon et al. 2007, p. 943).

The reference is actually the last IPCC report (at that time).

There is a preliminary terminology to explore. For example, (as we have seen) statistician prefer to talk about "Climate Simulator" rather than "Model" and *this is because the world "model" is heavily overloaded, as noun, a verb and an adjective* Rougier and Beven (2013). The model simulator is something that maps forcing into weather and then is the work of the statistician to convert such weather in climate.

7.2 Definition of Weather and Climate

We may have defined climate and weather several times but the new suggestion from the statisticians is particularly fascinating. The World Meteorological Organization defines climate as the average weather over the period 1961–1990. During the Newton Institute, this definition was somewhat reversed and so climate was defined as the process that generates weather.

The question posed at the meeting was how to define the climate in Exeter on 1 Jan 2070 (Stephenson et al. 2012) and the most accepted answer was "the probability distribution of weather in Exeter on 1 Jan 2070." The climate corresponds to a probability distribution including the extreme events. The weather is then just a random draw from this probability distribution. This means that the old saying "climate is what you expect" that could be translated as *climate is the expectation of weather* is no longer strictly valid. Climate can rather be modeled as a space-time stochastic process whose samples are the observable weather.

At this point, the question remains how to obtain weather and climate from a model. One possibility is to assume that climate is a stationary process so that the weather is identically distributed on all 31 days of January. On the other hand, climate in the same location could be defined using the weather observations of nearby locations close to the days of interest. This simply to emphasize that to define probability distribution of weather from weather observations requires a statistical model.

Before continuing with the discussion of what is weather and climate in statistical terms we need to introduce the concept of ergodicity. Although we have given several definitions for the climate we still have to describe a practical way to define it and a useful concept is that of phase space. In the early 1990s Barry Saltzman was trying to study glaciations with simple models. The starting point was to characterize the state of the system with just few variables. For example, the simplest model described the system with just two variables: the south latitude limit for the ice and the temperature of the ocean. Those two variables could change in time and for each instant the could be characterized by just one point with the vertical coordinate the temperature and horizontal coordinate the ice latitude. This representation describes the climatic system in a ice latitude—temperature space that is called phase space. This is a very familiar concept even for a first-year physics student and the most common example is the motion of a pendulum that in the phase space position-velocity gives a closed curve like an ellipse.

The exact characteristics of phase space are not known but we assume that the can visit all the points of phase space and the transition from one part of the phase space to another occurs smoothly. In principle, the climatic system can take any path in phase space and the time evolution of any of the path is governed by the physical laws. In order to apply this fact to the diagnostic of the observed and simulated climate, we have to assume that the climate system is **ergodic**. This term was invented by Boltzmann when he found a quantitative definition for the entropy of a gas something like proportional to the numbers of states that a gas could assume. In our case, ergodic means that the trajectory of the climate system will eventually visit all parts of phase space so that sampling in time is equivalent to sample different paths through phase space. The ergodicity of the climate system can be related to the property that climate is **transitive**. According to the original definition by Edward Lorenz the climate is transitive if the long-term average of its properties is the same independently of the initial conditions. This means that climate depends only on the boundary conditions like solar input, atmospheric composition, and so on. There is no proof that climate is transitive and to complete the Lorenz definition we could have an **intransitive** climate if the long-term average depends on the initial conditions. The same Lorenz recognized that a definition of climate based on long-term statistics does not make sense because it would not allows the climate to change and he then suggested that climate of our planet could be **almost intransitive**. Following the original Lorenz's definition (Lorenz 1968)

In an almost intransitive system, statistics taken over infinitely long time intervals are independent of initial conditions, but statistics taken over very long but finite intervals depends very much upon initial conditions. Alternatively, a particular solution extending over an infinite time interval will possess successive very long periods with markedly different sets of statistics.

This was the *Climate Determinism* paper of 1968. Then in 1975 Lorenz published another paper (Lorenz 1975) with possible examples of climate regimes that pointed out how the could be really almost intransitive, In the absence of changes in external forcing the climate exhibited change of considerable duration.

These concepts have relevance for the idea put forward on *Envirometrics* when it is affirmed that (Stephenson et al. 2012).

For an ergodic climate model with time-invariant boundary conditions, one may estimate the distribution of weather by taking samples from a single long simulation: the "distribution" representing climate is the invariant measure (state space attractor) for the system. However, climate forcings are not generally time-invariant, and so in practice we must be more subtle. To find the probability distribution of weather in Exeter on 1 January 2070, we could create replicates by considering an ensemble of simulations out to 2070 starting at random initial times during a period when boundary conditions may be assumed to be time-invariant (e.g., pre-industrial). Such an approach also assumes that the climate model is a perfect representation of the real and that the climate model is ergodic. Because climate models are imperfect representations of the real climate, calibration of the climate model runs is generally required, for example, bias adjustment of the means. Moreover, real future forcings are of course unknown and so the climate estimated using this approach is conditional on the forcing "scenarios" that are used.

It is rather interesting that the basic assumption is that the climate be ergodic otherwise the long simulation with random initial conditions does not assure basically the same climate. In climate prediction in essence, a meteorological model is used with different boundary conditions to obtain prediction several decades into the future. Climate modelers are assuming without asking questions that the is ergodic and then transitive. However in the latest years this Lorenz nomenclature is no longer used while the ergodicism of the climate system is left most to theoretical people that seem to be interested more in dignifying the science of climate. However just looking a Fig. 6.1 it does not seems that models in IPCC are ergodic. They start from different initial conditions and do not converge on the same climate.

7.3 How Statisticians Evaluate Models Results

A very important contribution of statistics could be the evaluation of model results. Whatever is the way models are evaluated data must have a central role. Peter Guttorp in the mentioned paper examines different questions related to the use of data the most important being homogeneization and comparison of database. The first issue is related to data adjustment when the station changes location or instrumentation. The procedure of homogeneization is important also in connection with the fact that what is now used ad climate data were usually taken for other reasons. The other important issue has to do with database and the homogeneization procedure is going to influence the database to the point that updating a database may influence the temperature by 0.5 C. Another points raised by Guttorp has to do with rating that is the newspaper announcements about the hottest or coolest year. There is a quite sophisticated way to determine the probability that a year is the warmest of coolest year.

When it comes to compare the models with data the first problem is the data from models. As already shown in Fig. 6.1 the range of temperature reproduced by the models has a spread of about 2 C for the so called historical period. Actually, the picture one sees on the newspapers is quite different and looks more like Fig. 7.1. This kind of figure (taken from Leroy et al. 2015) actually plots the deviations from the means obtained by simply averaging the data produced by the different models. Actually considering that climate data are obtained from weather data it would me more correct to plot long stretches of the distributions of the two type of data rather than anomalies. Guttorp has done that using probabilistic plotting that compare the observed data with models. That was done for two periods, the first one goes from 1930 to 1959 that is 30 years when the global warming "could be neglected" and the other from 1970 to 1999 during which the temperature could be increasing. The data used were those of the NASA Goddard Institute (GISS) while the model were the cumulative results of the CMIP5 experiment and those of two specific model NCAR and Hadley center (HadCM3). When compared with the earlier period the NCAR model had the best fit while for the warming period the opposite was true with HadCM3 giving the best fit. In any case, both models overestimated the warming. Guttorp asked at this point if both the data and the CMIP5 indicated a change between the two time period and the answer was clearly positive from the sophisticated technique he used. A major work has been done on the model evaluation by the group of statistician that participated to the "Envirometrics" meeting. The effort was concentrated on the so called Multi Model Ensemble (MME) simulations. Ensembles are created when in a single model the initial conditions or the physical parameters are perturbed. On the other end, MME are produced from a set of different climate models. The results of Fig. 7.1 are obtained by the MME used by the IPPC.

In their paper, Stephenson and collegues noted how

Despite their increasing complexity and seductive realism, it is important to remember that climate models are not the real world. Climate models are numerical approximations to fluid dynamical equations forced by parameterisations of physical and unresolved sub-grid scale processes. Climate models are inadequate in a rich diversity of ways, but it is the hope that these physically motivated models can still inform us about various aspects of future observable climate. A major challenge is how we should use climate models to construct credible probabilistic forecasts of future climate.

Now climate models do not produce probabilities so that the role of the statistician is to develop methods and models to arrive at such probabilities. A preliminary question that not many ask is why even for the historical data simulation models gives a dispersion as the one shown in Fig. 6.1. The main reason for that is the fact that one given the initial conditions it takes some time for the model to speed up and the initial average temperature produced may be different for the different models. As we have said many times, models are not to be confused with scientific theories and they are more like engineering application of some theory. Remaining in this framework

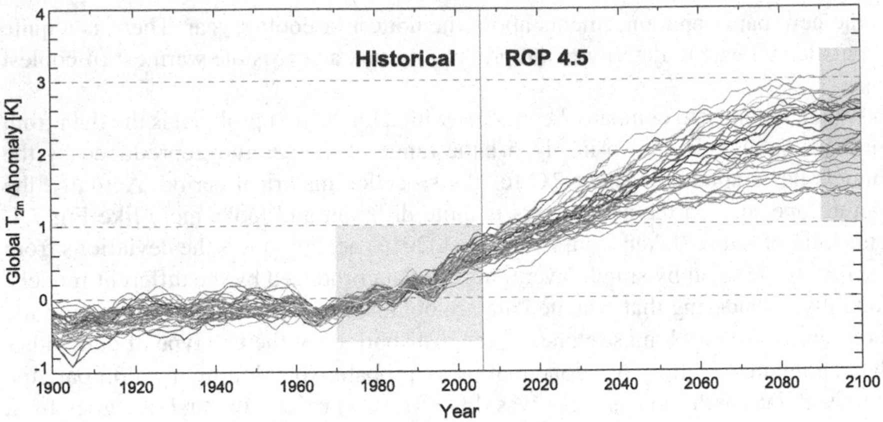

Fig. 7.1 The anomalies of temperature including the historical period (in the *beige box*). Figure is from Leroy et al. 2015 with Reference in Chap. 6

imagine now to apply fluid dynamics theory to design a wing of an airplane using different computer models. Similarly to what happens in climate modeling, you get different results like a 15% difference in the lift to drag ratio for the wing. It would be natural for the designing team to think that something is wrong with the models if not with the theory. They would go to the wind tunnel and test the design against the data and probably choose the model that better fits the data.

In climate modeling, people do not follow the same reasoning. They just make an average and present results as deviations from the mean, and not only that but they accept these results as reasonable. In this case, people have the experimental data which is the global average of the measured historical values of temperature so they could choose the model that best fits those data. On the contrary, they use the same group of models (there are about 40 now) to predict future climate and instead of choosing the model that best fit the data they develops sophisticated ways to average the models results (that we will see in a while).

There is still an additional problem with using average values. Only recently, some data have been published on spatial average compared to prediction that we mentioned in Chap. 4 (Mote et al. 2016). In particular, the results of the weather@home, a regional climate modeling group which is part of the climateprediction.net. These were obtained with almost 140,000 simulated years on a 25 Km resolution and referred to the western part of the US and among other things simulated years between 1960 and 2009. The published temperature maps may be interpreted visually through the difference or by calculating a correlation coefficient. The latter for the temperature is very good about 0.98 while the simulated temperature were cooler than the observed of about 0.66 C with the largest difference being about 2 C. The same simulations for the snow water equivalent does not look so good. In any case for this historical period, the difference of 0.66 C is comparable with the average change of temperature in the same period of time. Similar data always referring to the US were published in the NOAA Climate Change Web Portal. In this case, the period compared was much longer from 1911 to 2005 and the simulations were those of CMIP 5. Again major difference was observed in the mountainous regions (with the model being cooler by a maximum of 3 C) and in the northern regions where the model was warmer of roughly 3 C in the Arctic. These results confirm that although models gives roughly the right picture the detail are quite different and this confirms that policies based on regional simulation may be overconfident of the results.

Statistician suggests different ways to make averages and MME can be interpreted in two classical approaches, the frequentist interpretation and the Bayesian interpretation. The frequentist approach requires that the MME simulations can be regarded as a sample from a *notional population of "climate" models*. However to interpret probabilities in terms of some underlying population, it is necessary to understand the characteristics of such population. Statistics requires that the frequency interpretation is possible only if the elements are independent (i.e., in fair coin flips the events are independent). This raises all kind of questions regarding models how they originates and how they are developed. We all know that very rarely models are developed from scratch and we all know that models use so called packages from different sources. In light of that, it is rather humoristic that some people compare model development to biological development (Stephenson et al. 2012)

In other words, the development process is evolutionary whereby successive versions of a climate model (species) are subject to gradual change and occasional mutation, with the weakest species dying out over time. Each new species of climate model is likely to be more similar to currently existing species than if the set of viable species were sampled completely randomly each time. In addition, models are not developed in isolation and there is frequent communication between development groups. So rather than a random sample, it is likely that MMEs represent a relatively small, and probably relatively homogeneous, sample from the population of viable models.

In the same paper, Stephenson et al. conclude that MME outputs are not independent while in a previous paper Knutti (2008) found that the sample variance of the MME does not decrease as the reciprocal of the ensemble size. Another confirmation that frequency interpretation cannot be applied in this case.

There are different approaches when using the Bayes statistics. We mentioned the most obvious that is the Bayes model averaging when the models are averaged with a weight evaluated with a Bayes factor. Most recent methods use the output of MME and observations and are based on the initial work by Tebaldi et al. (2012). The basic assumption of this method is that the distribution of temperature (or other variables) of the MME is Gaussian and it is centered on the observed value. The Bayes theorem is used to evaluate the posterior distribution of the climate variables. A similar method has been proposed by Furrer et al. (2007) where the difference between the predicted variable and the same variable in the current climate is assumed to be linearly related with the error and the large scale signal. These approaches have been used to evaluate the regional response of MME and compared with the distribution of temperature or precipitation given by the model. They usually suffer from the lack of data both from the models and the observations.

Another, most radical way to use Bayes theorem has been previously illustrated in paragraph 6.5. This approach by Huang, Leroy and Goody shows that the "truth" can be reached within several decades by some kind of correction process of the model prediction with the data.

All these different proposals have a common objective and that is the reduction of the uncertainties and the outlook in this respect is rather depressing as it is well illustrated in the paper by Tebaldi and Knutti (2007)

For the decision-making process, it is important to know whether uncertainty in the evolution of future climate will remain at a similar level or whether it will be reduced substantially in the next decades. This uncertainty depends on the uncertainties in the emission scenarios, caused by uncertainties in social, economical and technical development, as well as uncertainties in climate model projections for a given scenario, caused by our incomplete understanding of the and the ability to describe it in a reasonably efficient computational model. While a probabilistic picture of climate model uncertainty is evolving (as demonstrated by many papers in this issue), emission scenarios so far do not have likelihoods attached to them. For these reasons it seems unlikely at this stage that projection uncertainty will decrease significantly in the very near future even on a global scale, and this is even more true for local projections. In principle, increased computational capacities combined with accurate long-term observations have the potential to substantially reduce the climate model uncertainties on the long run. However, even if the perfect climate model did exist, any projection is always conditional on the scenario considered. Given the often non-rational and unpredictable behaviour of humans, their decisions and the difficulty in describing human behaviour and economics in models, the perfect climate forecast (as opposed to a projection that is conditional on the scenario) is a goal that will probably be impossible due to the uncertainties in emission scenarios and the feedback loops involving the agents that the forecast is directed towards.

7.4 The Fortune Teller Approach and Related Stories

An old cartoon from the *New Yorker* shows the fortune teller woman outside her place with the sign *Madame Olga, your future simulated by a computer*, This paragraph report something like that for the simulation of future climate. We have mentioned before the simulator approach to climate modeling, The same Jonathan Rougier introduced the concept of **emulator** that is defined as *statistical model for a simulator, allowing prediction of the simulator output at untried values of the parameters*. In practice, you have the data produced by a model on the present climate and then you develop a statistical method to predict based on these data (but not only that) what will happen in the future. There are a few attempts to this end and one of the most popular has been published in 2013 by Peter Stott of the Hadley Centre and colleagues (Stott et al. 2013). The method requires results from a GCM experiment when the carbon dioxide concentration is increased abruptly (for example $4xCO_2$). Another thing you need to know is radiative forcing as it changes year by year in your chosen scenario. Then for each year you may scale the climate response and get at the least the change in temperature. In principle, this method may avoid to run huge and expensive GCMs. Stott and colleagues used two scenarios (Called Representative Concentration Pathways) RCP4.5 and RCP8.5. The first one implies a slow growth in the emission up to 2035 with a peak production of about 40 Gt/year of CO_2 and then a faster reduction to reach a stabilization at 5 Gt/year of CO_2. On the other hand, RCP8.5 has a steady growth with a stabilization at more than 100 Gt/year of CO_2 by the 2100. The baseline for the initial warming measured by the experimental data include the years 1986–2005. The emulated warming for the years 2020–2029 gives a range of 0.35–0.82 K for the RCP4.5 scenario while the CMIP5 simulation gives 0.48–1.00 K. For the RCP8.5 the emulator gives a range of 0.45–0.93 K against a CMIP5 range of 0.51–1.16 K. The conclusion they reach is that the upper range indicated by the CMIP5 simulation is inconsistent with the past warming. In practice, the CMIP5 simulation is too hot by a few tenths of a degree. The reason we called this the fortune teller approach is quite evident now. First of all look at the numbers as they give ridiculous second digit approximation that nobody is going to believe. Besides they use six models from CMIP5 (out of over 40) to develop their emulator and then use the results to criticize the MME from which they start. Another dubious element of this approach is the radiative forcing that is not determined experimentally but as always depends on the radiative transfer code that is used.

A more sophisticated approach to the emulator has been developed by Stefano Castruccio and coworkers Castruccio et al. (2014), Fyfe et al. (2016) of the Department of Statistics at the University of Chicago and by Ben Kravitz, McMartin and Kravitz (2016) of the Pacific Northwest National Laboratories(PNNL).

There is however a more serious proof that such emulator approach may be wrong and that is the work of Stephen Leroy from Harvard and his collegues (Leroy et al. 2015). Leroy intention was to evaluate the priorities according to the observation types being considered for a space based observing system. This idea was already in the above-mentioned work of Huang, Leroy and Goody but here it is extended to include the hindcast for the period 1970–2005. Notice that again they do not use observed data. The correlation maps between hindcast and climate prediction for the period 2090–2100 occasionally shows values greater than 0.5 that according to the authors should impliy a possibility to reduce the uncertainty by 30%. This correlation however happens for very limited geographical regions.

A very similar method has been used by Reto Knutti when he compared the present day's climate simulation by the CMIP3 with the prediction of the climate for the end of this century. He choose four models out of the 25 that at that time (2008) were part of the CMIP3 that showed bias as large as plus or minus 6 C. However even the model that reproduced most accurately the present climate did not show any correlation with the future climate. According to Knutti, this means that the accuracy showed by some model in reproducing the current climate is not a guarantee for the accuracy in predicting future climate. This is clearly seen in a paper published on Nature in 2002 where they had a picture similar to the Fig. 7.1. You can see that the historical period (in beige) shows a quite limited dispersion of the simulation while the difference between models increase for both the past and future climate. This may be the results of the effort by the modelers to reproduce the present climate even using some trick but also is the results of showing deviations from the average of all models. Another interesting point made by Knutti is the fact that the bias of a singular model is not very much different from the bias of several models and this may the evidence that models are not really independent (as we already said).

This brings us to a very recent paper by Ed Hawkins and Rowan Sutton of the University of Reading (Hawkins and Sutton 2016). They show that the reference period from which the deviations are calculated is quite important in determining the anomalies. This is clearly shown in Fig. 7.2. The upper part of this figure shows the simulations from the CMIP5 models for the historical period.

1861–2005 while the lower panels show the anomalies calculated when two different reference periods are chosen 1979–88 and 1996–2005. The colored lines are the results of a number of reanalysis and the according to the choice of the reference period the historical simulations either overestimate the warming (1979–88) or underestimate it (1996–2005). Not only that because the reference periods are going to influence also the prediction and choosing a more recent reference will reduce

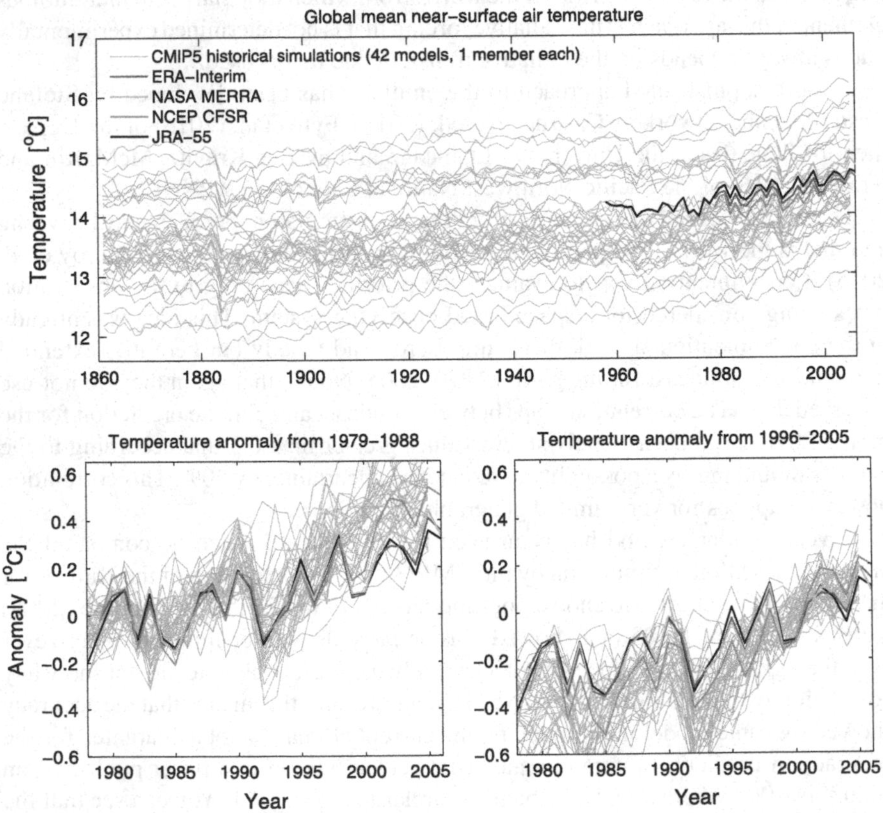

Fig. 7.2 The effect of the reference period on the global mean anomaly (see text for details)

the spread in the prediction of future climate up to about 0.5 C. The authors also confirmed the lack of correlation between the present temperature and the predicted one for the last 20 years of the present century as shown in Fig. 5.1. The reference period follows the recommendation of the World Meteorological Organization which dictate 30 years of average. We have seen that this is not exactly what the statisticians recommend.

It looks like again and again that the modeling community is interested mainly to give predictions for the policymakers, that is, those who are foraging their activity but they are not interested in doing science (whatever it is). The emulator example and the discussion how averages are made are illuminating. The latter in particular shows that even the basic concepts in the physical sciences are not clear even to their practioners (Fig. 7.3).

It is quite evident the total lack of correlation between the global temperature in the period 1979–2008 and the change due to the increase of greenhouse gases.

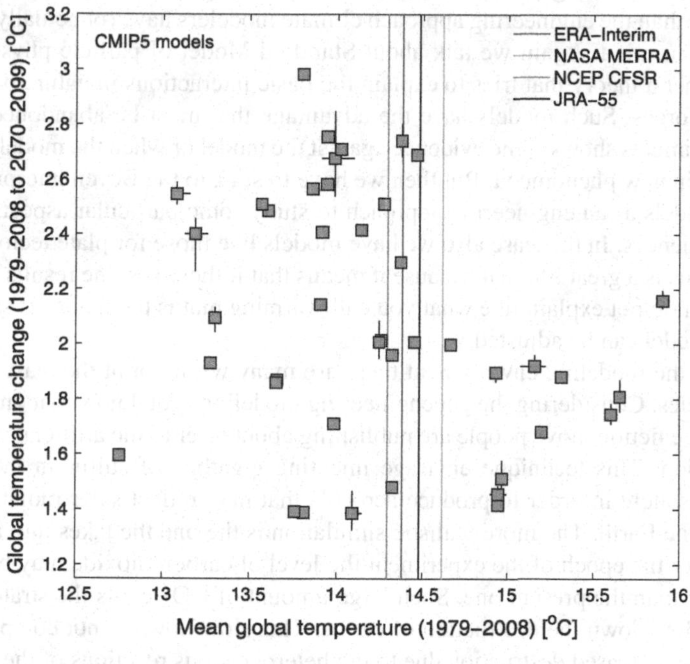

Fig. 7.3 The correlation between the predicted global temperature change by each of the 42 CMIP5 models and the present day temperature. The error bars on some model refer to the range of variation for that model. The *colored vertical lines* are the average for the same data of Fig. 7.2

Climate and Statistics

A climate scientist and a humanist at the end of each chapter will discuss what we intended to say and what they have understood. The humanist in this case should be quite disoriented.

H: One of the problems raised by the statistician is to change the "climate model" in "climate simulator." I would like to express a complete agreement with such idea. This reminds me of an old book by Percy Bridgeman, *The Logic of Modern Physics*, where he had a paragraph in Chap. 2 named *Models and Constructs*. His idea of models was quite different because he thinks about a scaled atom made by electrons which orbit the nucleus and his only worry is that after the scaling he can still talk about Euclidean space. The same confusion exist in the paper by G.E.P. Box (Box, 1979) in which he makes the statement "All models are wrong but few are useful." He refers models as ways in which a systems work.

C: Yes you are right. The concept of model in physics refers more to theory rather than the engineering approach climate modelers have (or actually do not have) in mind. Again we talk about Standard Model of particle physics that is rather a theory that tries to explain the basic interactions in nature based on four forces. Such models have the advantage that must be abandoned when experiments shows some evidence against the model or when the model cannot explain new phenomena. But then we have to stick to the Beven interpretation of models as an engineering approach to study some particular aspects of the geosciences. In this case also we have models like those for plate tectonics.

H: This is a great lifeboat because it means that if there is some results that the model do not explain like what you call warming hiatus the theory is save and the model can be adjusted.

C: In the modeling environment there are many wrong or at the least curious episodes. Considering the geoengineering modeling. You know that one of the science fiction move people are publishing about refer to the artificial volcanic eruption. This technique envisage injecting gigatons of sulfur in the lower stratosphere in order to produce aerosols that may reflect solar radiation and cool the Earth. The more realistic simulation is the one the takes into account that for the epoch of the experiment the level of carbon dioxide may be much larger than the present one. Such large amount of CO_2 cools the stratosphere and slow down the destruction of ozone. This slow down is not compensated for the increased destruction due to the heterogeneous reactions of the aerosol so that the most realistic simulation gives a large decrease of total ozone with a subsequent increase of the UV radiation at the ground. Now there are only a couple of paper published that show this effects while there are other papers that find completely different results with ozone increasing by roughly 1% . One must asks how these papers that are completely wrong/because do not even represent a science fiction reality got published in the first place.

H: another example is the warming hiatus where the modeling approach can be used to introduce ad hoc corrections to the hypothesis. The net results seems to say that the modeling does neglect some point but the as we understand it is correct. This will go on until the next discrepancy
C: The misunderstanding is basically that climate models are not to be confused with the climate theory which is what most of the modelers think. I remember an old saying that someone think the real nature are models
H: I was very much surprised to learn that the results we see on the newspapers are actually manipulated because the original results show a much larger spread. Even more surprising is the fact that you cannot choose the model that best fits the present data because this does not assures that the same model will give the most reliable results in the future.
C: The reason why simulations are not shown as they originally appears is that they make a bad and evident impressions of reproducing reality in such different ways. In case the comparison with engineering is maintained the best numerical algorithm is the one that best simulate the reality. Considering the dispersion of the results they do not give any reliability about the small numbers they claim on the prediction. An the you also learn as a great discovery that the period you choose as reference is going to influence the entity of the prediction and you still use the second decimal digit.
H: You probably refer to what in the chapteris called the fortune teller approach when you talk about the reliability of the today best model about future prediction. This may be not the worst thing because there are other evident problems considering that the future is seen with a total deterministic approach while there are factors that you cannot just predict on the political and the economical side. Beside the performances of statistics in the economy are of such poor standards that they promise to do any better for the climate.
C: You are right. It does not make any sense to attach significance to a forcing calculated to the second decimal digit when there are order of magnitude potential changes in the scenarios. I think prediction and statistics could give a very vague idea of what will happen in future and the reliability is going to decrease as the region where predictions are made is restricted. My choice would be simpler models a plenty of community involvement in planning and adapt for future possible climate

References

Box, G. E. P. (1979). Robustness in the strategy of scientific model building. In R. L. Launer & G. N. Wilkinson (Eds.), *Robustness in statistics*. Academic Press.

Castruccio, S., McInerney, D. J., Stein, M. L., Liu, F., Jacob, R. L., & Moyer, E. J. (2014). Statistical emulation of climate model projections based on precomputed GCM runs. *Journal of Climate*, *27*, 1829–44.

Furrer, R., Sain, S. R., Nychka, D., & Meehl, G. A. (2007). Multivariate Bayesian analysis of atmosphere ocean general circulation models. *Environmental and Ecological Statistics*, *14*, 249–66.

Fyfe, J. C., Meehl, G. A., England, M. H., Mann, M. E., Santer, B. D., Flato, G. M., et al. (2016). Making sense of the early-2000s warming slowdown. *Nature Climate Change*, *6*, 224–28.

Guttorp, P. (2014). Statistics and climate. *Annuals Review of Statistics and Its Application*, *1*, 7–101. Reproduced with permission of Annual Review © by Annual Reviews, https://doi.org/www.annualreviews.org/

Hawkins, E., & Sutton, D. (2016). Connecting climate model projections of global temperature change with the real world. *Bulletin of the American Meteorological Society*, *97*, 963–90 © American Meteorological Society. Used with permission.

Knutti, R. (2008). Should we believe model predictions of future climate change? *Philosophical Transactions of the Royal Society A*, *366*, 4647–64.

Leroy, S., Redaelli, G., & Grassi, B.(2015). Prioritizing data for improving the multidecadal predictive capability of atmospheric models. *Journal of Climate*, *28*, 5077–5090. © American Meteorological Society. Used society. Used with permission.

Lorenz, E. N. (1975). *Climatic predictability. Garp Publication Series*, *16*, 132–136.

Lorenz, E. N. (1968). *Climatic determinism. Meteorological Monographs*, *8*, 1–5 © American Meteorological Society. Used with permission.

MacMartin, D., & Kravitz, B. S. (2016). Dynamic climate emulators for solar geoengineering. *Atmospheric Chemistry and Physics*, *16*, 15789–99.

Maynard Smith, J. (1986). Molecules are not enough. *London Review of Books*, *8*, 8–9.

Rougier, J., & Goldstein, M. (2014). Climate simulators and climate projections. *Annuals Review of Statistics and Its Application*, *1*, 103–23.

Rougier, J., & Beven, K. J. (2013). Model and data limitations: the sources and implications of epistmic uncertainties. In J. Rougier, S. Sparks, & L. Hill (Eds.), *Risk and uncertainty assessment in natural hazards*. Cambridge University Press.

Solomon, S., Qin, D., Manning, M., Chen, Z., Marquis, M. (2007). Intergovernmental Panel on Climate Change, Climate Change: The Physical Science Basis. Contribution of Working Group I to the Fourth Report, Cambridge University Press.

Stephenson, D. B., Collins, M., Rougier, J. C., & Chandler, R. E. (2012). Statistical problems in the probabilistic prediction of climate change. *Environmetrics*, *23*, 364–72. Permission granted by Wiley material.

Stott, P., Good, P., Jones, G., Gillett, N., & Hawkins, E. (2013). The upper end of climate model temperature projections is inconsistent with past warming. *Environmental Research Letters*, *8*, 014024.

Tebaldi, C., & Knutti, R. (2007). The use of the multi-model ensemble in probabilistic climate projections. *Philosophical Transactions of the Royal Society A*, *365*, 2053–2075 by permission of the Royal Society.

Tebaldi, C., Smith, R. L., Nychka, D., & Mearns, L. O. (2012). Quantifying uncertainty in projections of regional climate change: A bayesian approach to the analysis of multimodel ensembles. *Journal of Climate*, *18*, 1524–40.

Chapter 8
Recent Developments

8.1 Introduction

Climate science as any other "scientific" endeavor manifest itself through publications and a simple statistics of them gives a faithful idea of what are the most discussed topics. There are two hot and recent development in the climate sciences which are the warming hiatus and the geoengineering. We will deal at length with them in the chapter but just to summarize, the warming hiatus is an apparent pause in the global warming which started at the end of last century and has been going on since then. Geoengineering is the purposeful modification of the environment to correct the effects of the anthropogenic activity. Both of these fields are quite controversial and they have produced a large number of publications. A statistics prepared in 2015 at the University of Texas, Austin uses the three key words, pause, slowdown or hiatus to recover that the total number of publications from 1990 to 2015 amount to 213 concentrated between 2013 and 2015. In the same time span, the citations reached a peak of roughly 850. Belter and Seidel (2013) published a bibliometric analysis on geoengineering examing the papers published between 1988 and 2011 for a total of 750 articles. It is interesting to note that half of these papers were published after 2008 and about 30.

30% of them were not research articles (i.e. reviews). The peak of more than 120 articles were published in 2011. The most pouplar topics ranged between solar radiation management, carbon sequestration (we will discuss if this technology could be considered geoengineering) and ocean fertilization. Also interesting is the fact that most of these papers came from the western world (dear old capitalism) proving that this is mainly an academic activity that follows the path traced by one of the earliest paper on the subject (Keith, 1992).

Most of the papers on the hiatus are in the direction of deny it as does the editorial in Nature on May 4 2017 but the problem is not that of fitting with ad hoc hypotheses the data. The problem is that the models which are predicting the future climate (in this case the short term climate) are not adequate to predict the details. As a matter of fact, the episode shows that those models are lacking the data that could prove

G. Visconti, *Problems, Philosophy and Politics of Climate Science*,
Springer Climate, https://doi.org/10.1007/978-3-319-65669-4_8

their predictions and those data will be never available because they refer to years to come.

As for the geoengineering, there are other problems. 30% of non research papers tells a story of lack of original research in this field. There is no statistics on the character of those paper whether they are experimental papers or theoretical (simulation papers) but we could assert that the percentage is well below 5%. So geoengineering research so far is simply a collection of papers that simulate the effect of the different technology with a majority between solar radiation management (SRM) and ocean fertilization. This is the technology which has received most of the experimental support and protesters.

The origin of the papers, mostly from the US and Europe, is a proof that the scientific community in these countries has found a way to gather public money to write different chapters of an infinite and poor science fiction saga.

8.2 The Warming Hiatus

In the last IPCC report (Hartmann et al. 2013), there is the official notification that something did not fit the GCM prediction about future climate .. *the rate of warming over the past 15 years (1998–2012; 0.05C [-0.05C to +0.15C] per decade), which begins with a strong El Nino, is smaller than the rate calculated since 1951 (1951–2012; 0.12C [0.08–0.14C] per decade)*. The possible slowdown in the warming was suggested earlier by Susan Solomon and co-workers (Solomon et al. 2010). The idea, based on some very noisy and incomplete data, was that following a decrease of stratospheric water vapor after 2000 also the forcing due to this gas would decrease slowing down the rate of temperature increase.

The reaction of the climate community and (the modeling in particular) has been quite atypical but it was a good occasion to publish all kinds of papers on the subject. Instead of catching the core of the question (models cannot predict the details of the future), the community opted for the ad hoc explanation with very few exceptions. Most recently, *Nature* published a couple of papers on the subject trying to defuse the problem (Risbey and Lewandowsky (2017), Medhaug et al. (2017). Most of the published papers ascribe the slow down to the variability of the system while others try to pinpoint the possible causes. For example Gavin Schmidt and collegues of the NASA Goddard Space Science Institute (Schmidt et al. 2014) make a list of possible errors made by the models which include the non-perfect mixing of greenhouse gases, the tropospheric and stratospheric aerosols, the solar irradiance, A similar approach has been used by Markus Huber and Knutti (2014) with the addition of the ENSO contribution. Only in a recent paper signed among others by Gerald Meehl (Fyfe et al. 2016) the hiatus is taken seriously and one of the authors (Shang-Ping Xie 2016) claim that the main possible cause of he hiatus is the only partially understanding of the energy budget of the Earth.

The hiatus spurred great interest in planetary energy uptake and redistribution in the ocean. A rigorous test of energy theory requires sustained global observations of planetary energy budget and ocean measurements beyond a depth of 2000 m. A challenge is to reconcile ocean heat redistribution and the regional sea surface temperature modes that cause hiatus events. From the growing body of hiatus research, tropical Pacific decadal variability has emerged as an important pacemaker of global climate, but its mechanism remains to be elucidated. In the central equatorial Pacific, the surface cooling during the hiatus rides above a subsurface warming highlighting the importance of ocean dynamics. Much as weather forecasting improved over time through daily verification, the early-2000s hiatus provides a valuable case study to test our observations, understanding and models of the.

Earlier explanations (Kosaka and Xie (2013) also attributed the hiatus to a redistribution of the heat where central was the role of the ocean. Again we are back to the problem that to predict climate, we need to predict the deep oceanic circulation that as we have seen before it is still very hard to predict. However, nobody has observed or remembered a very simple rule of the physical science that each model or theory must be tested against data. In the case of warming, hiatus models have failed to predict what happened and this should be enough to make people think. The different reasons that have been indicated as the cause of the problem could be included to correct the model and the prediction but considering the number of parameters that enter in a simulation such approach does guarantees nothing.

The hiatus was already popular among the negationists of the warming a few years ago. Just to illustrate the status of the climate science, it is worth to report a document by Richard Lindzen published on his web site (Lindzen 2008). This document follows an exchange between Lindzen and Stefan Rahmstorf a Professor of the physics of the oceans at Potsdam university. The debate appeared on the book edited by Ernesto Zedillo (Lindzen (2008), Rahmstorf (2008)) and was amusing for the low level of personal consideration. Rahmstord affirmed

When I was confronted with the polemic presented by Lindzen (in this volume) my first reaction was a sense of disbelief. Does Lindzen really think that current models overestimate the observed global warming sixfold? Can he really believe that climate sensitivity is below 0.5 °C, *despite all the studies on climate sensitivity concluding the opposite, and that a berely correlating cloud of data from one station, as he present in Figure 2–3, somehow proves his view? Does he onestly think that global warming stopped in 1998? Can Lindzen seriously beleive that a vast conspiracy of thousands of climatologists worldwide is misleading the public for personal gain? All this seems completely out of touch wit the world of climate science as I know it and, to be frank, simply ludicrous.*

As a young physicist working on aspect of general relativity theory, I was confrionted with a professor from a aneighboring university who claimde in newspaper articles that relativity theory was complete nonsense and that a cospiracy of physicists was hiding this truth from the public to avoid emarassement and cuts in their funding: (he refferred to the "Emperor's new clothes," as does Lindzen). The "climate skeptics" remind me of the "relativity skeptics," and perhaps the existence of people with rather eccentric ideas is not surprising, given the wonderful variety of people. What I find much harder to understand is the disproportionate attention and space that are afforded to such views in the political word and the media.

Rahmstorf it seems very much interested in the reactions from the newspapers and also to remind to his collegues his noble origins as general relativity theorist. The answer from Lindzen was adequate.

This finally brings us to Rahmstorf's absurd, pompous and pretentious association of one the landmarks in modern intellectual history, Einstein's General Theory of Relativity, with the primitive and crude world of climate modeling. Einstein was pained for much of his life by the fact that his general theory had a single adjustable parameter (the so-called cosmological constant). One can only imagine how he might have felt about this theory being compared with climate models that have almost an uncountable number of adjustable parameters

So much for the 'scientific level' of the climate science. But the story of the hiatus had another tail as narrated by Cornwall and Voosen (2017). The story start when Thomas Karl of NOAAA published a paper in *Science* in June 2015 that claimed the hiatus derived from a misinterpretation of the data. John Bates a former NOAA top brass accused Karl of having used data from NOAA that were not certified. The idea that formed in the media was that the paper by Karl et al. (2015) was published in time to influence the Paris Climate talk to be held in Paris the following December. Lamar Smith Chair on the House Committee on Science, Space and Technology, even called a testimony from Thomas Karl and hailed to the courage of John Bates. So one more time climate science was unable to disentangle from politics. Tim Palmer had noticed some problems with the climate politics in an interview by the New Scientist in May 2008. He noticed that politicians assume science to be a "done deal" and expressed doubts about the validity of the regional climate prediction of the IPCC. In case of a failure of such prediction, the entire reputation of the IPCC would be at stake.

In practice, again the message is: yes, warming is real but for now just neglect the details.

8.3 Geoengineering

Geoengineering has no noble origins although one of the most ancient traces could be found in the omnicomprehensive book by Svante Arrehenius *The making of the worlds* (1908) where he suggests to burn fossil fuels to prevent the next ice age. To Arrehenius, most people attribute the paternity of the Greenhouse effect. But the real origin of the concept goes to the science fiction with the novel *Star Maker* published in 1937 by Olaf Stapledon. Five years later the science fiction, novelist Jack Williamson conied the term *Terraforming* to indicate the reshaping of an entire planet to make it habitable. This concept was extended in a in 1992 by Martin Fogg in the book *Terraforming* which is now available on line. In this book, Fogg distinguishes between planetary engineering and geoengineering with the first as a technology to change the physical parameter of a planet while the second refers specifically to Earth.

Geoengineering is the intentional modification of the environment to balance the warming effects of increasing GHG. The most popular idea is called Solar Radiation Management that goes to influence the way solar radiation is absorbed by the Earth–Atmosphere system and one way to do that is by injecting "volcanic" amount of sulfur dioxide into the stratosphere. Volcanic means comparable to the amount produced by catastrophic volcanic eruptions of the order of 10 million tons of SO_2. Some people, however, assume that geoengineering is more an academic invention to draw some money and up to now it has been mostly the subject of model simulation. As we have seen the is quite poorly known and the GCM are engineering tools used mainly for prediction, it is quite unusual that the same models are used to study the effects of correcting the warming with fanciful means. The so-called sulfate engineering is the oldest way suggested to correct the problem by Mikhail Budyko in the 1970s and Wallace Broecker in the mid-1980s well before the suggestion widely reported in the literature by Paul Crutzen in the 1990s.

It is worth to note that two of the main supporter of geoengineering, Alan Robock of Rutgers and David Keith of Harvard (one of the few which is working in the lab) criticized the approach many years ago. Robock wrote a paper (Robock 2008) in which indicated 20 reasons why geoengineering was a bad idea while David Keith wrote a paper Keith and Dowlatabi (1992) in unsuspected 1992 on EOS (house organ of the American Geophysical Union) in which he suggested a *more coherent (though not large) research program in order to define fallback options needed to make reasonable policy choices.*

There are however many other people who believe geoengineering is the only way to moderate the warming caused by the GHG gas buildup. They apparently assume that the chances to reach a global agreement on emission reduction are dim. One of the most strong supporters of this approach is the same David Keith. It seems he is a strong advocate to carry out limited tests for the technique and in any case to proceed to a limited field deployment. In a recent paper, he concludes (SG is for solar engineering).

We expect that a central challenge will be balancing the breadth of approaches to be studied against the depth with which individual technologies and deployment scenarios are to be examined. While the GeoMIP G1 scenario is, for example, a poor proxy for policy-relevant deployment of SG, it has the advantage of being thoroughly examined by many researchers. To be policy-relevant and effective, the technical community needs to focus most of its effort on analyzing the efficacy and risks of a small set of technologies and deployment scenarios that show the most promise. To conclude, research over the last decade has explored a wide range of scenarios, but we argue that too much of that effort was devoted to a scenario choice that was suited to testing the performance of SG as a substitute for emissions reductions, rather than to more policy-relevant questions. Emissions cuts are a central part of any sane climate policy. Research on the science and technology of solar geoengineering needs embrace a design question: can a choice of technology, deployment and monitoring scenario be found that allows a significant reduction in climate risk? We propose-as a research hypothesis-that a moderate deployment scenario and reasonable choice of technologies can allow solar geoengineering to substantially reduce climate risks.

Unfortunately, the research he mentions is mainly based on modeling and we have a rather strange situation where the same models are used to predict the future climate and to "correct" the same predictions by adding some additional mechanism. The risks are to propagate also errors.

Before going to details, it is worth to remember a few historical episodes for geoengineering. In a paper published in 1991, (Cicerone et al. 1991) suggested that a way to reduce the ozone hole was to inject thousands of tons of propane or ethane in the polar stratosphere. The hydrocarbons would immobilize part of the chlorine atoms as hydrochloric acid and reduce the ozone destruction. A simple calculation indicated that should be "sufficient" an amount of 60,000 tons of propane to do the job. However, the requirement would be to inject that amount of hydrocarbons above 15 km and at that time (as it is today) there was plane capable to deliver such large amount of chemicals. A few years later, this would result the minor problem because the proponents had neglected an important heterogeneous reaction between the hydrochloric acid (HCl) and hypochlorous acid (HOCl). This reaction flourished out of the popularity of the newly rediscovered heterogeneous chemistry and so the same group (with some addition) published another paper (Elliott et al. 1994) which not only added the new reaction but had the bad surprise to find out that the effect was to accelerate the loss. Finally, the idea was wrong and this gives another example how sloppy is the reputation of climate science. In any other field (today this is very popular in bioscience) people withdraw their papers but here instead of withdrawing the earliest paper you write another one to show that the idea was wrong with good pace of the referees and the journal. We will see that there are few other examples of this kind.

The last historical items have to do with the so-called GeoMIP project. The acronym stands for geoengineering Model Intercomparison Project and if you visit their site you discover that is a very serious thing.

Geoengineering cannot be taken lightly. It would involve a level of planetary manipulation heretofore unseen. The implications are awe inspiring, and the consequences have the potential to be severe. geoengineering is fraught with unknowns that cannot be alleviated without coordinated study. GeoMIP endeavors to fill the need for such a study by prescribing certain experiments which will be performed by all participating climate models. Through this project, we will be able to ascertain commonalities and differences between model results of the climate response to geoengineering.

The things to be noticed is that they talk about experiments when there is no trace of experimental results. These are even less the classical GCM evidence (i.e., hindcast) so you will only find numerical grinding and results are listed to the second decimal digit. There are several group participating and presumably they are funded by state agencies. The accident of the previous episode will very unlikely repeat in GeoMIP because they need to compare the results among them but not with experimental data.

In what follows, we will talk mainly about SRM and we will make some mention to the most recent issue of Negative Emission Technology (NET).

8.4 Solar Radiation Management

The so-called Solar Radiation Management (SRM) envisages to reduce the amount of solar radiation reaching the lower atmosphere and surface of the Earth. As we mentioned in Chap. 2 what was clear to Jim Hansen in 1978 (Hansen, 1978) is quite dubious today. The original suggestion to inject aerosol into the stratosphere was made by the genius of Mikhail Budyko, the Russian climatologist, in 1976 in a book in Russian that was translated in 1977 (Budyko 1977) by the American Geophysical Union (AGU). However, when you read accounts of SRM the preferred reference is always the paper by Crutzen (2006) may be as a deference to the Nobel prize. Similarly to Budyko and Broecker the estimate of Crutzen were based on downscaling so that is much more interesting to report the original Budyko estimates. Based on the measured optical thickness of natural occurring aerosol he estimates a mass injection of 0.6 Mt of sulfuric acid. Such a mass require about 0.4 Mt of sulfur dioxide and then just 0.2 tons of sulfur. Because the lifetime of the aerosol particles is about 2 years the Budyko estimates is a rather optimistic 0.1 Mt of sulfur per year. As we know the latest estimates are at the least ten times as much but of course Budyko had very approximate data on the Agung eruption of 1963 and above all could not imagine that the effect of volcanic eruption is much less of what has been envisaged.

As we mentioned earlier in the book, there are recent papers by Canty et al. 2013, Fujiwara et al. (2015) and Wunderlich and Mitchell (2017). The latter uses nine reanalysis data sets and the analysis is concentrated to the eruptions of Agung (1963), El Chichon (1982) and Pinatubo (1991). The results show a clear warming signal in the lower stratosphere of about 1–2 °C while the signal in troposphere is quite weak and in some case completely absent at the surface. As for Canty et al. (2013).

If the AMV index is detrended using anthropogenic radiative forcing of climate, we find that surface cooling attributed to Mt. Pinatubo, using the Hadley Centre/University of East Anglia surface temperature record, maximizes at 0.14 °C *globally and* 0.32 °C *over land. These values are about a factor of 2 less than found when the AMV index is neglected in the model and quite a bit lower than the canonical* 0.5 °C *cooling usually attributed to Pinatubo.*

The magic effect found by Hansen et al. (1978) for Agung has disappeared and nobody seems to care in the literature. As we said before, in the biomedical field sometimes you find withdrawal of papers the same never happens in this field which has numerous and dangerous occasions. The signal that was assumed by all the geoengineering community is buried in the noise made up by the North Atlantic Oscillation, the Quasi Biennal Oscillation, ENSO, and so on. This results have obvious implications for the geoengineering as Canty et al. affirms.

If the factor of 2 reduction of volcanic cooling suggested here is borne out by future studies, the implications for geo- engineering of climate by artificial enhancement of stratospheric sulfate are immense. It would be straightforward to recalibrate Pinatubo as a proxy for geoengineering. Of greater concern is the fidelity of atmospheric General Circulation Models (GCMs) used to assess the response of the atmosphere to major volcanoes as well as geoengineering. Accurate GCM representation of the response to either perturbation requires realistic treatment of a myriad of physical processes, including the trapping of longwave thermal radiation by stratospheric sulfate aerosols, the dynamical response of the stratosphere, and changes in stratospheric ozone. The tendency of the GCMs used by IPCC (2007) to collapse the tropopause following major perturbations to SOD (Sulfate Optical Depth), which is not borne out by the ERA-40 reanalysis, suggests the physical response to stratospheric sulfate aerosols injection is not properly represented in modern climate models.

As usual, the geoenegineering community which is not accustomed to the experimental data has mostly ignored these results as have done the referees.

This is not the only pitfall of sulfate geoengineering and an important one is the acid deposition. Most of the calculations just assume a flux of SO_2 which is converted in the lower stratosphere in sulfuric acid H_2SO_4 which is the main constituent of the aerosol when diluted in water. Slowly (in about two years) these particles reach the troposphere and then are dissipated either as rain or as dry deposition contributing then to the acidification of the precipitation. Data from large, historical eruptions (Sigl et al. 2014) show that the distribution of the acid deposition is not uniform. However, simulation made by Toohey et al. (2013) show that middle latitude experience about 2% of the total mass flux, so it is possible to show that the sulfate deposition caused by the geoengineering approach produce fluxes of SO_4/m^2 comparable to those which now interest most of the US territory. From the chart provided by the National Atmospheric Deposition Program (NADP) (http://nadp.isws.illinois.edu) it is possible to note that only the north eastern region have deposition fluxes larger than those produced by the sulfate geoengineering.

The cost of sulfate geoengineering is another problem. We will start with some simple consideration on the implications of the large amount of sulfur required using a baseline of 1 Mt (10^9 kg) of SO_2. Our reference is a quite modern facility in Gujarat (India) (see https://www.chemtech-online.com/CP/Liquid_dec12.html) that can produce up to 500 kg/h of SO_2. Such facility would produce 1 Mt of sulfur dioxide in about 228 years or the same quantity of chemical could be obtained by building new 228 facility for each Mt. Considering only the energetic cost the same facility consumes 30 kwh/ton that will give a yearly consumption for 1 Mt of about 108 GJ/year or 10^8 BTU/year. The production cost of sulfur dioxide is never mentioned in the few papers on this matter that rather concentrate on the means to deliver the material up in stratosphere (McClellan et al. 2012) or on the uncertainties related to the exact values of climate sensitivity (Arino et al. 2016).

As for the means to deliver megatons of sulfur dioxide between 14 and 20 km, there has been a historical misunderstanding since the early papers on the subject (National Research Council 1992). The basic requirements are that the material must be injected into the stratosphere otherwise its residence time may be just too short. At the present time, there not planes or other platforms to do that so it is necessary to develop new planes. In this case, the development of a new plane capable of delivering 1 Mt could be around 1–2 billion dollars that could reach 5–8 billion for a 5 Mt delivery. The operational costs could reach about 1 billion/year with acquisition cost for the aircraft of 10 billion. Considering th past experience with military planes such cost may be within a factor of 10 not to mention the development time.

It is interesting to note that Broecker in his 1987 book (Broecker 1987) estimated a cost of 15 billion dollar per year just to manufacture the chemical (either SO_2 of sulfuric acid) and another 15 billion to deliver it. This would represent roughly 10% of the US defense budget of the time. The latest evaluation by Yosuke Arino and co-workers (Arino et al. 2016) estimated a cost oscillating between 2.5 and 5.9 trillion dollars that is 200 times as much. This would obtain a reduction of 0.5°. This represents roughly ten times the defense spending of the US in 2015 and about 5.5 the total budget of the US government. Even considering that this money should be spread among nations still it is very large amount of money and this brings us to the last point which cannot be resolved in a few lines: only rich countries could afford geoengineering and this will create some problem. More and more it looks like SRM is not only costly but full of uncertainties because up to this point has been just a computer exercise just like a play station analog.

8.5 SRM and Ozone

We have seen that SRM is essentially the injection of a considerable amount of aerosol in the lower stratosphere to mimic the effects of a huge volcanic eruption. Presumably, such practice would be implemented not before several decades from now and in any case when the GHG level would have reached higher level than today. One of the consequences of the increase of GHG would be the cooling of the stratosphere especially for the increasing level of CO_2 and the changing temperature would strongly affect the ozone chemistry. Banerjee et al. (2016) have shown what would be the cumulative effect on ozone of the projected changes in both GHG and Ozone Depleting Substances (ODS) without SRM. On the other hand there are a few papers on the SRM simulation (Tilmes et al. (2009), Jackman and Fleming (2014), Nowack et al. (2016)) which take into account the cooling and others (Heckendorn et al. (2009), Pitari et al. (2014),Weisenstein et al. (2015)) that just neglect it. In the presence of stratospheric cooling, there will be an ozone increase in the main (10 hPa) ozone layer. Cooling could also lead to more ozone loss in regions where PSC (Polar Stratospheric Clouds) formation and heterogeneous chemistry are important, e.g., the Antarctic lower stratosphere. The consequence is that those models that include cooling predict a stratospheric ozone increase while the others have an ozone decrease due to the effects of heterogeneous chemistry. Cooling brings about a slowing down of the chemistry that destroys ozone and this effect alone more than compensate for the effects of the heterogeneous chemistry. This is clearly seen in the Nowack et al. (2016) paper. The obvious question is how the community can accept without notice both results. By now we all know that environmental models are not representing reality but rather adhere to the principle of pragmatic realism (Beven, 2008). Even remaining in this limited perspective the publication of such different papers is hardly acceptable even though Weisenstein et al. (2015) state explicitly that

Thus our evaluation of ozone changes due to geoengineering by the injection of solid particles includes only chemical perturbations due to heterogeneous reactions on particle surfaces and not due to changes in temperature or dynamics induced by the geoengineering.

In this particular case, the only justification is that the aerosol used are of different composition than the usual sulfate (see also Pope et al. (2012) and Keith et al. (2016)) but in the other cases mentioned, their publication remains puzzling. In particular, Pitari et al. report ozone changes of the order of 1 DU that considering the different approximation have no meaning at all. On the other hand, their table 6 shows always in the same context numbers to which is very hard to attribute some meaning when they list changes up to the third decimal digit in radiative forcing and UVB that mean absolutely nothing. They mention to include a RCP4.5 scenario but do not explain how this influences their results. The remaining paper is by Heckendorn et al. who actually run a case of doubled CO_2 but apparently only to change the surface temperature. This paper is very much concerned about changing chlorine scenarios, and neglects the larger effect that would be introduced by a cooler stratosphere. Also to notice the discussion on the water vapor content based on the temperature changes that neglect those eventually introduced by the increasing CO_2 concentration. The general conclusion is that the peer review system is some case does not work although some of these papers have been published in journals where the reviews are public. The problem addressed by these papers is quite clear. In a word with changing greenhouse gas (GHG) content geoengineering is applied to correct the warming introduced by the increasing concentrations. The resulting simulations must take into account all the important consequences of the increasing GHG otherwise, publication of these results presents an incomplete picture. It is well known that aerosol particles affect the ozone and this problem has been extensively studied. If one transposes this problem to a different field (for example medicine) it would correspond to attempting to cure a patient for a negligible ailment and neglect the main disease. As shown recently by Ferraro et al. (2014) the increasing CO_2 does not affect only ozone but also the circulation of the stratosphere. Also Nowack et al. (2015) have shown that stratospheric ozone may have first order effects on climate sensitivity. The suspicion again is that either these simulations are the results of community project (like GeoMIP, geoengineering Model Intercomparison Project) or simply are reviewed by a quite restricted circle. This is very common practice in the field of modeling and not only in chemistry modeling. A typical example are the so-called negative papers where simulations fail the reproduce some experimental results and are published anyway without any critique of the model.

Alan Robock in his paper that celebrates the Budyko revival by Paul Crutzen (Robock 2016) tries to update the pro and con of SRM with respect to his earlier paper (Robock 2008) but he ends up with a table where benefits are summarized in 6 points against the 27 risks. Among the benefits are "unexpected benefits" and "beautiful red and yellow sunsets" so that the margin for advocating SRM is becoming very thin. Notice that risks and benefits are based on at the least 10 years (since Crutzen paper) of geoengineering studies which basically are computer exercise (the GeoMIP project) but again the presumption of scientists is without limits. Caldeira and Bala (2016) in the same celebration issue goes as far as

Most scientists see science in a largely Popperian framework [Popper, 1959], in which the essence of science is to construct falsifiable hypotheses and then try to falsify them. Physical scientists do not shy away from stating their personal opinions. While physical scientists such as Alan Robock and David Keith both try to generate and falsify hypotheses, neither shies from making their personal viewpoints known. Thus, David Keith writes books like "A Case for Climate Engineering" [Keith, 2013] and Alan Robock writes perspective pieces like "20 Reasons Why geoengineering may be a Bad Idea" (Robock 2008).

First of all, the definition of science is quite doubtful but geoengineering it is just the wrong example. This is an enterprise based only on simulation without any room for falsification. The data on the climatic effects of the volcanic eruptions seem to be negligible in the troposphere but still referees and authors do not seem to care. There are wrong papers in the literature which are not withdrawn and there are papers that neglect important effects like CO_2 cooling of the stratosphere and ozone. This is rather a confirmation that climate science, if it is a science, has a long way to go.

8.6 A Note on Negative Emission Technologies NET

Negative Emission Technologies (NET) indicates all the methods that can reduce the amounts of carbon dioxide in the atmosphere. They include forestation and afforestation up to the controversial BECCS (Bioenergy, combined with carbon capture and storage). In this case, bioenergy power plant would use agriculture residuals, organic waste and dedicated bionergy crops. The CO_2 produced by the plant would be in part refixed by the growing crops and in part captured and buried. The overall budget for carbon dioxide would be negative. The problem is that of these plants only one large demonstration example exists today (Anderson and Peters 2016). Nevertheless in the Paris agreements of 2015, the 1.5 C target could be reached only with massive use of such techniques. In very entertaining piece on *The Guardian* (Kruger et al. 2016), three expert notice that to achieve that goal by removing 600–800 billion tons of carbon dioxide in 15–20 years is not a practical option and so they recommend to exclude negative emissions from mitigation scenario. Rayner (2016) in a very amusing editorial compares the IPCC bureaucrat to the Azande people as narrated by Evans-Pritchard (1940). The Azande practiced to lodge a stone in a fork of a tree accompanied by a spell to delay the setting of the sun until they returned home from work. This was some kind of stimulus to hasten the work and magic in this case represent the determination to go home in time for dinner. IPCC use magic in its report (the NET) but then just does not work in the direction to promote work to reach that goals.

On the other hand, magic would really be necessary to implement the NET as described by Smith et al. (2015). The techniques for realizing negative emission are fundamentally the ones known as BECCS, DAC (Direct Air Capture), EW (Enhanced weathering), and AR (Afforestation–Reforestation). All these technologies (which for the sake of brevity are not illustrated here) need energy, cultivatable land, and water. For example, to remove about 12 Gt of CO_2 per year, BECCS would require from 380 to 700 Mha of land (Amazonia amounts to about 550 Mha), with an energy consumption of 170 EJ/year (exajoule/year equal to 1018 J/year): the world energy consumption comes to 600 EJ/year. However, the most important consideration is that until now these techniques have not been tested, and industrial systems do not exist. Thus, the Paris decision is based on a strong wishful thinking, rather than on a reality. This is really a geoengineering enterprise complete of its science fiction side.

Recent developments

A climate scientist and a humanist at the end of each chapter will discuss what we intended to say and what they have understood. The humanist in this case insist on the ethical implication especially of geoengineering

H: It is clear that most of this chapter just does not believe that neither geo-engineering of advanced mitigation technologies (NET) may do the job to limit global warming.

C: Although quite young the literature on the subject is immense. Some of these hypothesized technologies only correct part of the problem. The Solar Radiation Management implies that you correct only the warming but not all the other consequences of the carbon dioxide increasem like ocean acidity.
H: I would like to raise several question the main one is about the so-called ethics of geoengineering which has not been discussed very much so far. Again to me is a little bit exaggerate to talk about ethics or philosophy when the questions are rather technical for example who has the right to deploy a SRM system and what are the side effects.
C: I think questions are even more basic. You know that in 2015 the president of National Academy of Science, Marsha McNutt proposed to abandon the geoengineering term and substitute it with Climate Intervention. The reasons are that in this case there is no "geo" but rather "climate" and that engineering refers more to a established practice rather than hypothetical intervention.
H: Right! All these hypotheses look more like science fiction stuff of the early days and there is just nothing like engineering that would implies to have a blueprint for intervention. What is impressive is that 30
C: Here we have a clear case of proposals that do not have in most cases any experimental base and in many cases are just the same computer run of climate prediction with some parameters changed.
H: I would like to go back to the ethical issue. Most of the papers and discussions are in obscure academic style. It looks like more a justification to show their own erudition than a real challenges to the very serious questions climate intervention poses. A very important is related in my opinion to assure a long duration (someone says 10,000 years) to a SRM intervention (I remember this was raised by prof. Raymond Pierrehumbert). This may be unthinkable for a single state and is even less realistic for and international consortium
C: This clearly refer to the fact that while the effects of an artificial volcanic eruption manifests rapidly the same happens once SRM would be suspended. The cooling of the planet would be suddenly interrupted and this would have disruptive consequences.
H: Another problem is related to the so-called side effects which are not discussed at length
C: All right there are side effects and the reason why they are not discussed in the text is their uncertainty. Starting from the drop in temperature you have seen that most recent data show very serious doubts that would implies using larger amount of aerosol to be injected in the stratosphere. The other consequences of the large volcanic eruption are uncertain as well. As an example there are a few papers which document the precipitation decrease over land after the Pinatubo eruption.

If this effect is maintained also with sulfur injection that would be a great risk for agriculture. Beside that we discussed at length that one of the major uncertainty of the GCM or LAM (limited area models) concern the regional climate. So all the hypotheses about monsoon precipitation and similar stuff are very marginal indeed.
H: This is the result of working with predictions which are uncertain (those for global warming) to which further possible errors are added. This kind of consideration affects another worry about climate intervention and that is the disparity between rich countries and poor ones. Not only that! I know that Canada for example may protest against cooling the climate because a warming for them would implies benefit for agriculture.
C: The discussion about who gain and who loses from climatic change is much older than IPCC and it is very hard to disentangle. The available simulations are just unable to tell which region will experience what. So I think the discussions based on that are just useless.
H: Another interesting question is related to the political color of climate intervention. This is quite important because it decide whether the initiative is supported by conservative government and oil companies.
C: Well it looks like environmentalists are definitivily against geoengineering because it is configured ad a violence to natural processes. You may also remember that early proposals on "sunscreen for planet Earth" were formulated by Edward Teller and clearly the proposla is supported by negationist and oil company. On the other hand Bill Gates has offered some financial support to David Keith.
H: This has some important general consequences. If implementing a SRM system will make the reduction of GHG emission a less urgent matter and is very strange that one of the reasons used to justify SRM is that of buying time before government reach an agreement on the curbing of emissions.
C: As we already said SRM would leave untouched (and sometime even exacerbate) other environmental problem caused by the increased CO_2 concentration. The already measurable decrease in the Ph of the ocean will continue unabated as well as other problems related to the other greenhouse gases. In addition you must consider all the side effects produced by the SRM.
H: These discussions are really premature. It is like to discuss of the problems of using nuclear energy before having demonstrated the possibility to build a nuclear reactor. On the same side no discussions were made when the Genoma project was initiated that could have important ethical implications.
C: Some of the problems agitated are common to the climate debate. For example the different impact on the poor and rich countries and the conversion of fossil fuels based energy to renewable.

References

Anderson, K., & Peters, G. (2016). The trouble with negative emissions. *Science*, *354*, 182–183.

Arino, Y., Akimoto, K., Sano, F., Homma, T., Oda, J., & Tomoda, T. (2016). Estimating option values of solar radiation management assuming that climate sensitivity is uncertain. *PNAS*, *113*, 5886–5891.

Banerjee, A., et al. (2016). Drivers of changes in stratospheric and tropospheric ozone between year 2000 and 2100. *Atmospheric Chemistry and Physics*, *16*, 2727–2746.

Belter, C. W., & Seidel, D. J. (2013). A bibliometric analysis of climate engineering research. *WIREs Climate Change*, *4*, 417–427.

Broecker, W. S. (1987). *How to build a habitable planet*. Eldigio Press.

Budyko, M.I. (1977). *Climatic change*. AGU.

Canty, T., Mascioli, N. R., Smarte, M. D., & Salawitch, R. J. (2013). An empirical model of global climate-Part 1: A critical evaluation of volcanic cooling. *Atmospheric Chemistry and Physics*, *13*, 3997. License from Copernicus Publications.

Caldeira, K., & Bala, G. (2016). Reflecting on 50 years of geoengineering research. *Earth's Future*, *4*, 10–17. Reprinted by permission from Wiley Copyright and permission.

Cicerone, R. J., Elliott, S., & Turco, R. P. (1991). Reduced Antarctic ozone depletions in a model with hydrocarbon injections. *Science*, *254*, 1191–1994.

Cornwall, W., & Voosen, P. (2017). How a culture clash at NOAA led to a flap over a high profile warming pause study. Science Podcast. February 8.

Crutzen, P. J. (2006). Albedo ehnancement by stratospheric sulfur injcetions: a contribution to resolve a policy dilemma. *Climatic Change*, *77*, 211–219.

Elliott, S., Cicerone, R. J., Turco, R. P., Drdla, K., & Tabazadeh, A. (1994). Influence of the heterogeneous reaction HCl + HOCl on an ozone hole model with hydrocarbon additions. *Journal of Geophysical Research*, *99*, 3497–3508.

Ferraro, A. J., Charlton-Perez, A. J., & Highwood, E. J. (2014). Stratospheric dynamics and mid-latitude jets under geoengineering with space mirrors and sulfate and titania aerosols. *Journal of Geophysical Research*. doi:10.1002/2014JD022734.

Fujiwara, M., Hibino, T., Mehta, S. K., Gray, L., Mitchell, D., & Anstey, J. (2015). Global temperature response to the major volcanic eruptions in multiple reanalysis data sets. *Atmospheric Chemistry and Physics*, *15*, 13507–13518.

Hartmann, D. L., et al. (2013). Observations: Atmosphere and surface. In T. F. Stocker & G.-K. Qin (Eds.), *Climate change 2013: The physical science basis*. Press: Cambridge Univ.

Hansen, J. E., Wang, W. C., & Lacis, A. A. (1978). Mount Agung eruption provides test of a global climatic perturbation. *Science*, *199*, 1065.

Heckendorn, P., et al. (2009). The impact of geoengineering aerosols on stratospheric temperature and ozone. *Environmental Research Letters*, *4*, 045108.

Huber, M., & Knutti, R. (2014). N. *Nature Geoscience*, *7*, 651–656.

Jackman, C. H., & Fleming, E. L. (2014). Stratospheric ozone response to a solar irradiance reduction in a quadrupled CO2 environment. *Earth's Future*, *2*, 331–340.

Karl, T. R., et al. (2015). Possible artifacts of data biases in the recent global surface warming hiatus. *Science*, *348*, 1469–1472.

Keith, D. W., & Irvine, P. J. (2016). Solar geoengineering could substantially reduce climate risks–a research hypothesis for the next decade. *Earth's Future*, *4*, 549–559. Reprinted by permission from Wiley Copyright and permission.

Keith, D. W., & Dowlatabadi, H. (1992). A serious look at geoengineering. *EOS*, *73*, 289–293.

Keith, D. W., Weisenstein, D. K., Dykema, J. A., & Keutsch, F. N. (2016). Stratospheric solar geoengineering without ozone loss. *PNAS*, *113*, 14910–14914.

Kosaka, Y., & Xie, S. P. (2013). Recent global-warming hiatus tied to equatorial Pacific surface cooling. *Nature*, *501*, 403–407.

Kruger, T., Geden, O., & Rayner, S. (2016). Abandon hype in climate models. *The Guardian*. April 26.

Lindzen, R. S. (2008). Is the global warming alarm founded on fact? In E. Zelillo (Ed.), *Global Warming: looking beyond Kyoto*. Brookings Institution Press. Permission granted from Brookings Institution Press.

Lindzen, R. S. (2008). Response to Stefan Rahmstorf's "Anthropogenic Climate Change: Revisiting the Facts". www-eaps.mit.edu/faculty/Lindzen/L_R-Exchange.pdf

McClellan, J., Keith, D. W., & Apt, J. (2012). Cost analysis of stratospheric albedo modification delivery systems. *Environmental Research Letters*, *7*. doi:10.1088/1748-9326/7/3/034019.

Medhaug, I., Stolpe, M. B., Fischer, E. M., & Knutti, R. (2017). Reconciling controversies about the 'global warming hiatus'. *Nature*, *545*, 41–47.

National Research Council. (1992). *Policy implications of greenhouse warming: mitigation, adaptation and the science base*(Chap. 28, pp. 433464). Washington, D. C.: Natl. Acad. Press.

Nowack, P. J., Abraham, N. L., Maycock, A. C., Braesicke, P., Gregory, J. M., Joshi, M. M., et al. (2015). Stratospheric ozone changes under solar geoengineering: implications for UV exposure and air quality. *Atmospheric Chemistry and Physics*, *16*, 4191–4203.

Pitari, G., et al. (2014). Stratospheric Ozone Response to sulphate geoengineering: results from the geoengineering Model Intercomparison Project (GeoMIP). *Journal of Geophysical Research*. doi:10.1002/2013JD020566.

Pritchard, E. E. (1940). *Witchcraft oracles, and magic among the azande*. Oxford: Clarendon Press.

Pope, F. D., Braesicke, P., Grainger, R. G., Kalberer, M., Watson, I. M., Davidson, P. J., et al. (2012). Stratospheric aerosol particles and solar radiation management. *Nature Climate Change*, *2*, 713–719.

Rahmstorf, S. (2008). Annthropogenic climate change: revisting the fact. In E. Zelillo (Ed.), *Global Warming: looking beyond Kyoto*. Brookings Institution Press.

Rayner, S. (2016). What might Evans-Pritchard have made of two degrees? *Anthropology Today*, *32*, 1–2.

Risbey, J. S., & Lewandowsky, S. (2017). The 'pause' unpacked. *Nature*, *545*, 37–39.

Robock, A. (2008). 20 reasons why geoengineering may be a bad idea. *Bulletin of the Atomic Scientists*, *64*, 14–18.

Robock, A. (2016). Albedo enhancement by stratospheric sulfur injections: More research needed. *Earth's Future*, *4*, 644–648.

Schmidt, G. A., Shindell, D. T., & Tsigaridis, K. (2014). Reconciling warming trends. *Nature Geoscience*, *7*, 158–160.

Sigl, M., McConnell, J. R., Toohey, M., Curran, M., Das, S., Edwards, R., et al. (2014). Insights from Antarctica on volcanic forcing during the common era. *Nature Climate Change*, *4*, 693–697.

Smith, P., et al. (2015). Biophysical and economic limits to negative CO_2 emissions. *Nature Climate Change*, *6*, 42–50.

Solomon, S., Rosenlof, K. H., Portmann, R. W., Daniel, J. S., Davis, S. M., Sanford, T. J., et al. (2010). Contributions of stratospheric water vapor to decadal changes in the rate of global warming. *Science*, *327*, 1219–1223.

Tilmes, S., Garcia, R. R., Kinnison, D. E., Gettelman, A., & Rasch, P. J. (2009). Impact of geoengineered aerosols on the troposphere and stratosphere. *Journal of Geophysical Research*, *114*(D12), 305.

Toohey, M., Kruger, K., & Timmreck, C. (2013). Volcanic sulfate deposition to Greenland and Antarctica: A modeling sensitivity study. *Journal of Geophysical Research*, *118*, 4788–4800.

Weisenstein, D. K., Keith, D. W., & Dykema, J. A. (2015). Solar geoengineering using solid aerosol in the stratosphere. *Atmospheric Chemistry and Physics*, *15*, 11835–11859. License from Copernicus Publications.

Wunderlich, F., & Mitchell, D.M. (2017).Revisiting the observed surface climate response to large volcanic eruptions. *Atmospheric Chemistry and Physics*, *17*, 485–499.

Xie, S.-P. (2016) The hiatus spurred great interest in planetary energy uptake and redistributionin the ocean. A rigorous test of energy theory requires, Nature Climate Change, *6*, 345–346. Reprinted by permission from Macmillan Publishers Ltd.

Chapter 9
Some Conclusion

9.1 Introduction

The following was the opening remarks that Harold Schiff (a famous atmospheric chemist) made in 1973 at the IAGA (International Association of Geomagnetism and Aeronomy) held in Kyoto (Dotto and Schiff 1978).

Once upon a time there lived in the land of IAGA, in the kingdom of **Aeronomy,** *strange creatures called aeronomers. Little was known about these creatures because they lived most of their lives in the remote area of the kingdom, more than 60 kiloleagues from the Earth.*

Not so long ago a part of their kingdom, known as **Stratos,** *was threatened by the invasion of a flock of big birds who make noise, that sounds something like—sst. Some of the creatures of* **Aeronomy** *rushed to* **Stratos** *to try to discover what these birds may be doing to their kingdom. Some came because they heard these birds could also lay golden eggs.*

We soon learned that there are three kinds of **aeronomers.** *There is group of high priests called* **modelers.** *They never go outside their temples where they try to prophesy what the big birds will do by examining the entrails of large animals called* **computers.** *Another group, who appear to be the worker drones called* **experimenters,** *spend most of their time in noisy, smelly rooms called* **laboratories** *playing with little boxes whose purpose seems to be the generation of random numbers called* **data.** *A strange relationship exists between the modelers priests and the* **experimenters.** *The priests feed the data to the computers animals and then study their entrails. They then tell the* **experimenters** *what kind of new data the animals need and the* **experimenters** *rush back to their* **laboratories** *and make more black boxes. The third group is called the* **observers.** *They also made black boxes but they throw their boxes into the sky. Most of the time the black boxes break. Sometime they too give* **data** *which the high priests also give to their animals. However the animals sometime get sick if they eat this data and may even die if too many different kinds of* **data** *are fed to them at the same time. However the high priests have become very clever at getting their animals to accept almost anything. The diet of these animals seem to lack one*

G. Visconti, *Problems, Philosophy and Politics of Climate Science*,
Springer Climate, https://doi.org/10.1007/978-3-319-65669-4_9

essential nutrient called transport data. Unfortunately, these data are grown mostly by **dynamicists** *who live in the land of* **Tropos.** *Only recently have the borders between* **Tropos** *and* **Stratos** *been opened to allow dynamicists and aeronomers to talk to each other*.

The 70s were the time when the debate was on the environmental consequences of the Supersonic Transport (SST) that is how a fleet of supersonic transport could affect the straospheric ozone. This humourous story needs only few adjustements to fit our discussion on models and data, although there are significative differences. The most comprehensive study on the effects of the SST was the CIAP (Climatic Impact Assessment Program) published in 1974 that could not predict the problem of ozone depletion caused by the chlorine chemistry. CIAP was mostly concentrated on the nitrogen oxides catalytic cycle and could not even imagine what could be going on in the Antarctic stratosphere. There was a tremendous lack of experimental data and this is something the ozone problem has in common with the climate problem. The conflict between models and data existed also at that time, although models are never evoked but rather constitute the entrails of large animals. What lacks in our case is the group of *experimenters* while the boxes throw in the sky by the observers could very well be the satellite instrumentation. The models used at that time had a big problem on transport which was parameterized using the eddy diffusion coefficients. Today the corresponding problems for GCM could be interactive inclusion of the ocean circulation so that contact must be encouraged between atmospheric circulation modelers and oceanic modelers including the processing of carbon through the biogeochemical cycles cycles.

Still there remain important differences to be noticed. Modeling the ozone depletion is much simpler than modeling the climate system, but the ozone case shows the great importance of the experimental data. Once the results of the dedicated measurement campaign were known it was clear that the culprit for the polar ozone depletion was the chlorine heterogeneous chemistry and all the dissenting voices disappeared rapidly. The other important differences with the present situation are of the emotional type. Ozone depletion was related to an increase of of the UV radiation which could imply an increase of skin cancer cases. Also the ozone hole was a dramatic and spectacular phenomena and it reached easily the imagination. The climatic change beside some occasional extreme weather event (of doubtful attribution) is a quite slow and ponderous process. The main point is that chemistry models were unable to predict anything like the ozone hole until it was discovered and the question could be if something similar is happening for the climate prediction considering the scarcity of available data.

This is the final chapter in which the reader should find some useful conclusion and we will offer a few. The most important one is the answer to the question on the existence and reality of climate science. Today climate science is identified with global warming and this does not appear right. It would be like to identify medicine with oncology neglecting all the other problems. The limits are even more severe considering that global warming is the only problem with a very limited amount of data except for the historical period. Being a very complex system, climate needs to be studied with long and costly experimental procedure where the spaceborne observations will have a growing role. Even if we limit our interest to Global Warming we have to find ways to assess the correctness of the prediction in the absence of a direct experimental evidence as it happens for the weather forecast.

Another possible conclusion to be discussed is what we need in order to convince governments to take the appropriate actions to mitigate or to adapt to the possible climate change. The question is if we need such detailed models or we may use simplified schemes which give useful keys to understand the problem. The road to more and more sophisticated models advocated by many implies the development of dedicated research centers. These should be complemented and integrated with appropriate observing system.

9.2 Again What Is Climate and Are GCM's Reproducing It?

The problem with the GCM concept is that it attempts an impossible task of seeking reality. In the first place reality (the state of the climate) cannot be defined objectively. This concept requires a little detour and to return again to the definition of climate. Despite all the effort produced by the statisticians again the definition contained in a paper by Edward Lorenz dated 1975 still holds (Lorenz 1975)

weather is often identified with the complete state of the atmosphere at a particular instant. As such the weather is continually changing. Weather prediction is then identified with the process of determining how the weather will change as time advances, and the problem of weather predictability becomes that of ascertaining whether such prediction is possible.

Climate may be identified with the set of statistics of an ensemble of many different states of the atmosphere. Particularly when the real atmosphere is replaced by an idealized mathematical system the ensemble is often taken to consist of all states during an infinite time span. In this event the climate, by definition does not change, and climatic prediction and predictability become meaningless.

At this point it seems that climatic change does not exist either but Lorenz defines what he intends for climate change

Our interest in what we call climatic change has arisen because atmospheric statistics taken over a rather long time span may differ considerably from those taken over a subsequent span....For our purposes we may therefore define climate in terms of the ensemble of all states during a long but finite time span. Climatic prediction then becomes the process of determining how these statistics will change as the beginning and end of time span advances, and climate predictability is concerned with whether such prediction is possible.

The most important point made is the distinction between weather and climate and in principle there is a way to do that if we look at the variance of a variable like temperature as shown in Fig. 9.1. Actually this figure that was originally drawn by Murray Mitchell in 1976 (Mitchell 1976) and updated by Ghil et al. (2002) is the

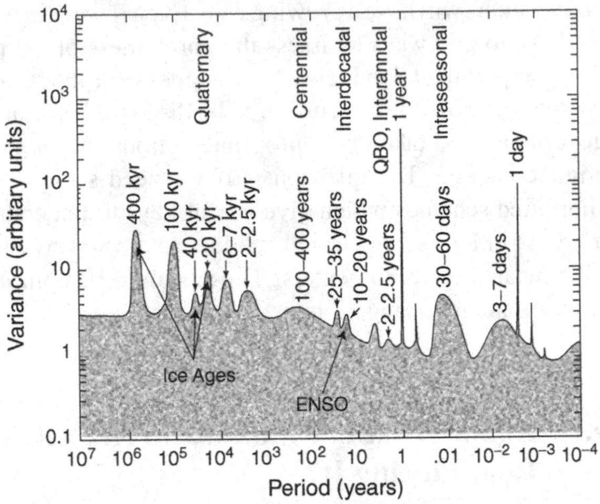

Fig. 9.1 The power spectra of climate variability over the last 10 Million years

power spectrum of the change of temperature recorded in the last 10 million years. Nobody has such a record so that the figure is rather a composite from different information. The original figure arrived down to the beginning of the solar system (roughly 5 billion years). This figure shows a few very sharp peaks (1 year, 1 day) which correspond to definite forcings: the diurnal cycle and the year cycles. But this suggest that also the other peaks must represent specific forcing.

For example the ones at 20, 40, and 400 kyear (1 kyear = 1000 year) represent the possible astronomical forcing of the ice ages (precession of the equinox, inclination and eccentricity). The fact that these peaks are not sharp is because the recurrent periods are not constant. This is quite evident in 10–20 year peak, which corresponds to the El Nino Souther Oscillation phenomena that has not a constant period. The intervals in which the spectrum is constant (for example for longer periods than 1 million year) corresponds to what is known in physics as stochastic forcing. As a matter of fact it can be shown that the climatic system had basically a red noise spectrun (the variance decreases as the frequency increases) going to high frequency (low value of the period) while has a white noise spectrum for very low frequency. This is quite evident in the figure and the better explanation for that was given by Klaus Hasselmnann (Hasselmann 1976). He assumed the weather to be some kind of noise for the and at the same time it constitutes the forcing. He then considers this forcing to be stochastic in the way the molecular motions force through the collisions the motion of larger particles (Brownian motion). Similarly to what happens in the brownian motion the slow components of the climatic systems (ocean, ice sheets, vegetation) acts as integrators of the forcing term producing a spectrum which is "red" and whose amplitude increases with the squate root of time. Hasselmann argues that in order to have a stationary situation (amplitude constant with time) a number of negative feedback mechanisms are necessary. The simple one is the temperature as it rises it emits more and more radiation until it stabilizes. This happens in the limit of low frequency when the forcing term is equilibrated by the emission of the surface and atmosphere.

However, in the actual record (that becomes more and more uncertain as we go back in the past) this low frequency limit is not observed but for longer time scale one incur in the tectonic forcing (i.e., climate change due to plate tectonics) and so you cannot define the climate as limit for low frequency. Of course, as we have seen we can pick up specific aspect of the climate like the ice ages, but we cannot define such episodes analitically.

So again what is climate precisely? Richard Goody says

We may think of societal answers to this question ("Times of Feast, Times of Famine" to repeat the famous title by LeRoi Ladurie's book) but not scientific in terms of well-defined and measurable quantities. This suggest some subjective definition of climate one based on human terms and the other in mathematical terms. One suggested possibility is to define climate based on different averaging intervals like 10, 20 or 30 years and to show that the conclusions reached differ e.g. the temperature rise over 100 years. If they do not differ then the constraints on the problem are more important than the averaging time employed.

This definition would be included in the Lorenz quasi intransitive argument (see Sect. 7.2) when he observed that in that case long term statistics may have the same solution while finite interval averaging depends very much on initial conditions. *Alternatively a particular solution extending over an infinite time interval will possess successive very long periods with markedly different sets of statistics.* An alternative definition of climate based on time interval of different length does not solve the problem.

It may be worth to recur again to Edward Lorenz when he defines prediction of the first kind for prediction of specific climate states. Like 1 or 2 years El Nino, while predictions of the second kind refer to predictions of climate statistics like global warming. Lorenz extends the concept of transitivity and intransitivity also to models as he affirms

It has been conjectured that the atmosphere-ocean-earth system is almost intransitive on a rather long scale. The two climates would be the glacial and interglacial climates, while the transitions from one climate to another would be presumably occupy but a small fraction of the total time. If we suppose momentarily that these transitions are brought about only by some catastrophic processes we can say that a numerical model which is correct except for omitting the catastrophes would be intransitive. Intransitive models may of course be much less elaborate. Some of the simplest climatic models are intransitive and in fact possess two steady state solutions resembling interglacial and glacial climates

This is necessary to justify his conclusions that *The more readily predictable the climate is in the first sense, the more difficult it is to predict in the second sense* And the explanation for that readily follows

Climatic predictability of the first kind would be enhanced by almost intransitivity, or by slowly varying features such as SST (Sea Surface temperature) patterns, which could lead to a relatively high probability that the coming month or year or centuries

would depart from the normal one in a known fashion. But to say that subsequent months or years or centuries are predictable is to say that they are not representative of the climate as determined from longer ensembles. Consequently, a numerical integration of one simulated month or year or century, as the case might be, would be insufficient for the purpose of estimating longer term statistics. To investigate predictability of the second kind, then our numerical integration should ideally extend beyon the range of predictability of the first kind

If we look at the bundle of GCM results for the simulation of historical data we see that the final results strongly depend on the initial conditions while all the models seem to reproduce the same trend. All of the Lorenz hypotheses refer to a stationary system while the GCM simulations include historical forcing terms. So the observed common trend seems to show simply that GCM describes a linear system, whose final results depend on the initial conditions and the system (or the model) seems to confirm to be definitively intransitive. The problem, however, remain whether GCM are reproducing correctly the reality or put in a different way if we can define "reality" in terms of GCM output.

This can only be justified in terms of pragmatic realism if all the parameters are recognized and all the processes make precise use of the established laws of physics and chemistry. Everyone knows that this is not the case. So what is the point of arguing one form of reality against the other? Again Richard Goody

Despite this skepticism we can see some value in the procedure but only if one recognize that the GCMs are heuristic models. Models that attempt to understand the relationships between variables based on the best representation possible of the physical processes. Many years ago this would be a 'grey' model of radiative equilibrium. This method is often used in science to understand what matters to which parameters without trying to solve an impossible problem. This is not a matter of proof or certainty but expert opinion, which is sometimes (often?) wrong. We have no doubt that the weight of believably on the question of anthropogenic global warming lies with the IPCC. But we do not think that they are predicting reality. It is then interesting to discuss whether simpler models may useful as well.

9.3 Long and Wasted Years?

Thirty-seven years ago, the National Research Council (1979) published what is known as Charney report entitled *carbon dioxide and Climate: A Scientific Assessment Report of an Ad Hoc Study Group on carbon dioxide and Climate*. The group of experts that prepared the report included Jule Charney, Chairman, Akio Arakawa, James Baker, Bert Bolin, Robert Dickinson, Richard Goody, Cecil Leith, Henry M. Stommel and Carl Wunsch. The report is around 35 pages and follows earlier report printed at the beginning of the 70s like the *Study of Man's Impact on Climate (SMIC)*.

If one reads the *Summary and Conclusions* of this document (barely 2 pages), one cannot notice that 37 years have passed without changing the basic conclusions.

We have examined the principal attempts to simulate the effects of increased atmospheric CO_2 *on climate. In doing so, we have limited our considerations to the direct climatic effects of steadily rising atmospheric concentrations of* CO_2 *and have assumed a rate of* CO_2 *increase that would lead to a doubling of airborne concentrations by some time in the first half of the twenty-first century. As indicated in Chap. 2 of this report, such a rate is consistent with observations of* CO_2 *increases in the recent past and with projections of its future sources and sinks. However, we have not examined anew the many uncertainties in these projections, such as their implicit assumptions with regard to the workings of the world economy and the role of the biosphere in the carbon cycle. These impose an uncertainty beyond that arising from our necessarily imperfect knowledge of the manifold and complex climatic system of the earth.*

At that time the doubling of CO_2 roughly mean reaching 680 ppm and at that time it was the most important greenhouse gas.

The primary effect of an increase of CO_2 *is to cause more absorption of thermal radiation from the earth's surface and thus to increase the air temperature in the troposphere. A strong positive feedback mechanism is the accompanying increase of moisture, which is an even more powerful absorber of terrestrial radiation. We have examined with care all known negative feedback mechanisms, such as increase in low or middle cloud amount, and have concluded that the oversimplifications and inaccuracies in the models are not likely to have vitiated the principal conclusion that there will be appreciable warming. The known negative feedback mechanisms can reduce the warming, but they do not appear to be so strong as the positive moisture feedback. We estimate the most probable global warming for a doubling of* CO_2 *to be near* $3°$ *with a probable error of* $\pm 1.5°$. *Our estimate is based primarily on our review of a series of calculations with three-dimensional models of the global atmospheric circulation, which is summarized in Chap. 4. We have also reviewed simpler models that appear to contain the main physical factors. These give qualitatively similar results.*

The report also recognized the role of the ocean

One of the major uncertainties has to do with the transfer of the increased heat into the oceans. It is well known that the oceans are a thermal regulator, warming the air in winter and cooling it in summer. The standard assumption has been that, while heat is transferred rapidly into a relatively thin, wellmixed surface layer of the ocean (averaging about 70 m in depth), the transfer into the deeper waters is so slow that the atmospheric temperature reaches effective equilibrium with the mixed layer in a decade or so. It seems to us quite possible that the capacity of the deeper oceans to absorb heat has been seriously underestimated, especially that of the intermediate waters of the subtropical gyres lying below the mixed layer and above the main thermocline. If this is so, warming will proceed at a slower rate until these intermediate waters are brought to a temperature at which they can no longer absorb heat.

The warming seems to be unavoidable although the details at that time were quite fuzzy

The warming will be accompanied by shifts in the geographical distributions of the various climatic elements such as temperature, rainfall, evaporation, and soil moisture. The evidence is that the variations in these anomalies with latitude, longitude, and season will be at least as great as the globally averaged changes themselves, and it would be misleading to predict regional climatic changes on the basis of global or zonal averages alone. Unfortunately, only gross globally and zonally averaged features of the present climate can now be reasonably well simulated. At present, we cannot simulate accurately the details of regional climate and thus cannot predict the locations and intensities of regional climate changes with confidence. This situation may be expected to improve gradually as greater scientific understanding is acquired and faster computers are built.

It is interesting to note that for a CO_2 doubling, the warming estimate were not too much different from the ones we have from the last IPCC report. As shown in the Chap. 4 of the document (Models and their validity) the estimates were based on two General Circulation Models, one from Syukuro Manabe and collaborators from the Geophysical Fluid Dynamics Laboratory in Princeton, and the other from Jim Hansen, NASA Goddard Institute. There were also simpler models like the radiative convective models. It is rather surprising that the committee made some specific observation about the models capability

Since, moreover, the snow-ice boundary is too far equatorward in H1 and too far poleward in M1 and M2 (see Appendix), we believe that the snow-ice albedo feedback has been overestimated in the H series and underestimated in M1 and M2.

We may conclude that about 40 years ago, the conclusions are not too much different from the present one including the gross feature of a greater warming at high latitude and considerable uncertainties in the regional prediction. Besides, it looks that the few models available at that time were accurately scrutinized a very much different approach from the IPCC. In this case, there is no attempt to compare the validity of the different models.

From the point of view of the average citizen during this period not very much has changed. The Charney report could be used by any politicians as a base for the energy policy of the future but nothing has been done. As a matter of fact this is mentioned in the foreword

The conclusions of this brief but intense investigation may be comforting to scientists but disturbing to policymakers. If carbon dioxide continues to increase, the study group finds no reason to doubt that climate changes will result and no reason to believe that these changes will be negligible. The conclusions of prior studies have been generally reaffirmed. However, the study group points out that the ocean, the great and ponderous flywheel of the global, may be expected to slow the course of observable climatic change. A wait-and-see policy may mean waiting until it is too late.

Only with the information contained in the Charney report any government could sign today protocols (starting from Kyoto) because the changes in some of the most important climate indicators are such to create enough worries. On the other hand, the same data could be used to predict details at regional level and would present the same degree of uncertainty of regional prediction based on more sophisticated GCMs.

A common shortcoming of both the Charney report and the late IPCC is the lack of experimental data. In the former there is no mention of possible proofs of global warming and the same problem remain for the IPCC. The first proof of the existence of the water vapor feedback has to wait ten years when Amee Raval and Venkateswaran Ramanathan (Raval and Ramanathan 1989) proved it. There exist the classical experimental proofs for the temperature increase and the rise of the ocean level although they remain quite uncertain. The possible confirmation of the behavior of cloud cover as predicted by the models came just in 2016 (Norris et al. 2016). There are many others possible proofs of the effects of global warming but clearly since 1979 there has been a lack of willingness to implement an experimental research program focused on the study of the physical and chemical processes that may explain the global warming as well as the verification of the predicted consequences.

9.4 Think About It

It is time now really to draw some conclusion. The most pressing one is about the existence of a science of climate. Most of the research effort in this field has been finalized and concentrated to reveal if the emissions and build up of the greenhouse gases has an influence on the climate and the main tools used to this end are General Circulation Models (GCM). GCM's must be regarded as an expression of what is called pragmatic realism. This philosophical (?) concept was introduced by Keith Beven (Beven 2002) but was detailed in his most recent book (Beven 2008). The idea is quite simple as reported in the 2002 paper

It is quite possible to develop and use environmental models without any explicit underlying philosophy, for discussions in different areas of environmental science). Many practitioners do, of course, although most might have the aim of developing and using models that are "as realistic as possible", given the constraints of current knowledge, computing capabilities, observational technologies and available time. This type of implicit or pragmatic realism seems quite natural and appears to be intrinsic to modeling efforts as diverse as the coupled General Circulation models (GCMs) being used to predict future climate change and modeling of the biogeochemical cycles processes

In his book Beven details the concept considering that in any case the environmental models (and so GCM's) are open systems, that is, they have uncontrolled boundary conditions that involve different kind of fluxes with the rest of the world. In principle such a systems cannot be solved but the "pragmatic realism" make the assumption that if you don't have anything better you must be happy with what you got.

There are many advantages in adopting the pragmatic realism stance as for example that the models could be improved constantly as long as you get new data and technological improvement. Again Beven

It could be argued from a pragmatic realistic perspective that the problems of defininig the parameterisation of a fornal model and defining effective parameter values in system characterisation are not, in themeselves, sufficient to require anew philosophy of environmental modelling but merely reflect the technological constraints of today (computing limitations, measurements limitations, theoeretical limitations, etc.). The expectations is that new technological developments will reduce the significance of today's problems.

Beven also clarifies how models must be intented as just tools .... *without making any realist claim.* and he mentions another philosophical position which he defines as *Instrumentalism*: *Instrumentalism is, if you like, the engineering view of modelling.*

At the center of the Instrumentalism view is that all the scientific theories of the past have proven to be false. This need not, however, prevent us from making statements about the nature of thing, including quantities that are not directly observable. Such statements, including models, can allow us to make useful predictions in practice, it is just that we should not believe they are real. This is the sense in which instrumentalism is anti-realist especially since it also allows that some statements about the nature of things may be subjective. It follows that the only justification for scientific theorising is empirical adequacy, a concept that is also intrinsic to the pragmatic realism outlined above

If we follow this concept, we arrive at the conclusion that GCMs are just tools useful to make predictions, but they do not represent the real working of nature. We have seen that even their predictions are subject to several limitations and some time need ad hoc corrections to make them consistent. Beven mentions that there are other theories identified as "pragmatic realism" even if he neglects the relevant contribution of Hilary Putnam.

Climate science is not simply the predictions of future temperature or precipitation because of the build up of GHG but must include many other topics like the explanation of ice ages, the origin of decadal variability, the El Nino prediction to mention a few. Today everything seems to be focused on global warming and this attitude has overestimated the weight of models with respect to other approaches. The planning of possible experiments to study the may involves several years of data gathering but it is a well-known practice in many other fields of science (for example, astroparticles physics) and should not be feared by scientist and administrators. Again this was the case of CLARREO and it seems obvious that the requirements for meteorology are quite different from those for climate. Rather than going to build immense computer center to fake a CERN for climate, we need to propose and plan ambitious experiments adequate for the very complex problems related of the climate system. As already noted by Lorenz a complete simulation of a glacial–interglacial cycle it not compatible with the fact that a GCM needs a day of computer time to simulate a year so that there is no hope to have such an essential simulation in a reasonable amount of time.

It is interesting to remember a paper by Richard Goody and co-workers published on the Bulletin of the American Meteorological Society, *Why monitor Climate* to ask for a long term monitoring program of climate (Goody et al. 2002)

A second reason for the collection of climate data is to support projections of future climates, taking both anthropogenic influences on future climate and the human impacts of future climate into account. The climate community claims to have predictive capability for climate of both the first and the second kind (Lorenz 1975). *Predictions of the first kind are for specific climate states, for example, for 1-or 2-year predictions of El Nino or similar oceanic phenomena. Predictions of the second kind are predictions of climate statistics, for example, for global warming 50 or 100 year into the future. The United States has made good progress with El Nino predictions and their consequences for regional climate, and the National Oceanic and Atmospheric Administration (NOAA) sees this as an operational responsibility. For this activity there is, again, a clear societal objective against which research and monitoring programs can be evaluated. The situation differs little in principle from weather forecasting. Numerical models give a detailed description of the evolution of El Nino, and observing systems can be devised to test these predictions. The Tropical ocean and Global Atmosphere (TOGA) array of fixed buoys in the Pacific has been a great success in this respect.*

And then going to the problem of global warming prediction

For predictions of the second kind, monitoring objectives are less clear. There are climate scientists who believe that useful predictions of the second kind are not possible, so that there is no point in devising observing programs to support them, and perhaps it may eventually turn out to be so. However, predictions of global warming 50,100 year from now are being made, most importantly by the United Nations Intergovernmental Panel on Climate Change (IPCC), and they are being acted upon (with some reservations) by many politicians. **It is to this development that climate science owes its present notoriety and much of its funding**. *The political issue is what actions can be taken consistent with uncertainties in the predictions; the scientific issue is not how to make longterm predictions, we have elaborate numerical models to do that, but how to increase the credibility and decrease the uncertainty of these* *climate predictions. This, then, should be a primary societal objective for climate monitoring and for climate research in general. It has not, in the past, occupied a central position in the U.S. climate research program.*

There are at the least two points to emphasize. On one side, the acknowledgement that notoriety of climate science is strictly related to the prediction of the global warming and the other the objective to reduce the uncertainties in the prediction. Unfortunately the planning of energy resources again must be made a few decades in advance and the goverments are more willing to act if they are stimutated by scientific and solid confirmation about the climatic consequences of the building up of GHG. Projects like CLARREO should have started at the time they were proposed (roughly the epoch of the BAMS paper) so that now we would have some definitive evidence about global warming. We have also seen that using Bayesian statistics the "truth" can be achieved after several decades improving the ensemble models prediction. All the evidence indicates that time necessary to reach a scientific conclusion can be of the order of 50 years of dedicated research with appropriate funding. During this time a "no regret" approach could be adopted by the nations of our planet that will also be a sign of their willingness to stick to the signed programs (or promises?).

We may then conclude that climate science is a respectful science, but it is not identified with GCMs that are rather engineering tools climate science must be involved in the study of all climate problems (not only global warming) using all the necessary tools, numerical and experimental with the community claiming adequate funding which are much more justifiable than other research. The impression now is that climate science is the same as global warming. As Richard Goody reccommends

We should separate the climate policy from the climate science. The former will use GCMs and make such interpretations as they wish. The latter should be asking why the climate is what it is, using heuristic models and scientific reasoning to look at one link at a time. If GCMs are used they should be used in an heuristic (i.e. experimental) mode.

Some Conclusion

A climate scientist and a humanist at the end of each chapter will discuss what we intended to say and what they have understood.
H: The conclusion that can be drawn leave very few hopes for a future based on proved scientific beliefs. I understand that considering the state of the GCM's predictions the government programs are based more on emotions rather than logical arguments. It looks that this is also a very cheap option because assuming a bleak future climate save a lot of money in planning meaningful experiments.
C: I did not thought about this side of the problem but you are right. It will take a few decades before we can be certain about the presence and the entity of the warming unless there is some unexpected events. As we have mentioned something similar happened for the ozone hole that was not predicted by the models at that time. The reaction times were quite fast and measurements campaign were organized and carried out in a few years and when the "smoking gun" was found the governments were almost forced to sign the Montreal protocol. However there is a basic difference between the two cases because the ozone hole was spectacular but unharmful while a climatic catastrophe would cost much more in economic and human terms.
H: I see another problem with funding which is the same mentioned somewhere in the book by Carl Wunsch and that is that the slow climate component is not consistent with the tenure system and/or the agency funding. This may go in the direction outlined by Harald Schiff and that is a quite strong link between modelers and funding. It much more cheap to fund model development (even a CERN for modeling) rather than a satellite project.
C: Yes that could be very well the case and this take us to another dangerous link between large computing centers and their government funding and the need to have results that could be please them.
H: This bring us to the link between GCM and reality. It looks like you must resort to philosophy and again to the "pragmatic realism" concept. It seems to me that you are trying to get an imprimatur for the modelers conclusions. The parallelism between engineering models and GCM does not look reasonable to me. You have a terrible spreading in your predictions and if this would happen in the design of and engineering project you would be obliged to go back and check if your theory was wrong.
C: You are quite right as a matter of fact even in the hydrological modelling which is were Keith Beven come from such a different results could produce unpleasant consequences. The "pragmatic realism" is useful because clarifies that GCM's are not the thoery of climate and rigourosly they are not even "models" in the sense William Bridgman describe in his *The Logic of Modern Physics* as we already mentioned. He makes the example for the model of an atom

The model may have many more properties than correspond to measurable properties of the atom, and in particular, the operations by which the space of the model is tested for its Euclidean character may (and as a matter of fact I believe do) not have any counterpart in operations which can be carried out on the atom. Further, we cannot attach any real significance to the statement that the space of the atom is Euclidean unless we can show that no model constructed in non-Euclidean space can reproduce the measurable properties of the atom. In spite of all this I believe that the model is a useful and indeed unescapable tool of thought, in that it enables us to think about the unfamiliar in terms of the familiar. There are, however, dangers in its use: it is the function of criticism to disclose these dangers, so that the tool may be used with confidence.
In practice in this case we have something quite different from the "pragmatic realism" because it is admitted that the model represents something that does not exist in nature while the main point is that the model reproduces all the properties of an atom. Beside we do not have different models which predict different behavior like in the case of GCMs.
H: Although I should be on a different side I think the simplification that present today predictions as the same of those of 37 years ago it is a too strong simplification.
C: Yes you may be right but I think this is the core of the problem. We know that an increase of GHG could bring basically a warming which is observed in the last 100 years or so. What we really don't know are the details of the consequences of such a warming. There is a general agreement that for example regional predictions (which are the most useful for planning) present a very large uncertainty. Consider that some time even the weather forecast may go wrong. Also it does not make sense to have 40 models that basically predict the same global behavior especially when you consider trends. The exaggeration is useful in pointing out the problems of the present predictions.
H: The section which elucitades the "pragmatic realism" is rather interesting from the philosophical point of view except for the fact that what is intended in philosophy for the same term is quite different. I think the most interesting part is that the concept leave the door open to possible improvement and new data.
C: I think that the main point is exactly the new data and the necessity to plan long term and complex experiments on the . The scientific climatic community should be not afraid of this endevour which is quite common in physical sciences. Think about the planning and execution of the LIGO experiment for the gravitational wave detection or other long term experiment in astroparticle physics. I think this aspect is quite well known to the community but is not appreciated by the government funding agencies.

H: The outcome could be dramatic because the consequemces of climate changes could manifest in a violent and unexpected manners for example through an intensification of extreme events. This would be an unequivocal sign the global warming is real and that the community would unprepared to mitigation or adaptation strategies.
C: I agree. The decision on the crucial experiments should have been taken several years ago in order to have now the proof that humans are changing the environment. On the other hand take example on how the problem has been solved for other pollulants like sulfur dioxide. The carbon dioxide problem remain too complex mainly from a political point of view.
H: Which bring us the geoenegineering question. I have the impression that this remain an academic invention to draw some research money. If I must take seriously the most important issues are about the authorities that should control even the simplest experiment.
C: I think you can make all the excercises based on non existent possibilities. The main problem is the way you delivery your million of tons of sulfur dioxide. At the present time there are no aircraft or other systems to do that and the cost and time to develop new ones is prohibitive. Beside the energy you consume in the process produces amount of carbon dioxide that constitues an appreciable percentage of the CO_2 your tryng to save.

References

Beven, K., (2008). *Environmental modelling: An uncertain future?*. CRC Press.
Beven, K. (2002). Towards a coherent philosophy for modelling the environment. *Proceedings of the Royal Society of London, Series A, 458*, 1–20 by permission of the Royal Society.
Dotto, L. & Schiff, H. (1978). *The ozone war*. Doubleday. Excerpt(s) from OZONE WAR by Lydia Dotto and Harold Schiff, copyright © 1978 by Lydia Dotto and Harold Schiff. Used by permission of Doubleday, an imprint of the Knopf Doubleday Publishing Group, a division of Penguin Random House LLC.
Ghil, M. (2002). Natural climate variability. In M. C. Mac Cracken & J. S. Perry (Eds.) *Encyclopedia of global environmental change*. Wiley. © Reprinted by permission from Wiley Copyright and permission.
Goody, R., Anderson, J., Karl, T., Miller, R. B., North, G., Simpson, J., et al. (2002). Why monitor the climate. *Bulletin of the American Meteorological Society, 83*, 873–78. © American Meteorological Society. Used with permission.
Hasselmann, K. (1976). Stochastic climate models part I theory. *Tellus, 28*, 473–85.
Lorenz, E. N. (1975). Climatic predictability. *Garp Publication Series, 16*, 132–36. Permission granted by WMO.
Mitchell, M. J. (1976). An overview of climate variability and its casual mechanisms. *Quaternary Research, 6*, 481–93.
NAS (1979). *Carbon dioxide and climate: A scientific assessment*. National Academy Press. Permission from National Academy Press.
National Research Council (1979). *Carbon dioxide and climate: A scientific assessment*. National Academy Press.
Norris, J. R., Allen, R. J., Evan, A. T., Zelinka, M. D., O'Dell, C. W., & Klein, S. A. (2016). Evidence for climate change in the satellite cloud record. *Nature, 536*, 72–5.
Raval, A., & Ramanathan, V. (1989). Observational determination of the greenhouse effect. *Nature, 342*, 758–61.

Appendix A

A.1 The Bayes Inference

The definition for inference can now be found everywhere (Wikipedia?). Bayesian inference is a method of statistical inference in which Bayes' theorem is used to update the probability for a hypothesis as more evidence or information becomes available. So that everything goes back to Bayes theorem that we have exposed in the text. Thomas Bayes was a Presbyterian minister lived in England in the eighteenth century (1702–1761). One of the most simple application of the Bayes inference has to do with establishing, based on the statistics of the flips, if a coin is fair. The inverse problem is the most interesting that is to establish something on the nature of the coin after 10 flips. Not only that but we can estimate a parameter which gives the degree of fairness of the coin and how this estimation can be improved with other 10 flips for example. This is the inference. To derive the Bayes formula, we may follow the beautiful book by Singh Sivia (Sivia and Skilling 2006). We start with a sample space where N microscopic events may happen. The different microscopic possibilities result in two events A and B, whose number N_A and N_B are proportional to the different areas of the space. The overlapping of the two areas represent the probability of the joint event N_{AB}. The probability of the different events is then

$$Pr(A) = N_A/N;\ Pr(B) = N_B/N;$$

$$P(A \mid B) = N_{AB}/N_B;\ P(B \mid A) = N_{AB}/N_A$$

The the probability Pr(A, B) that both A and B is given by

$$Pr(A, B) = N_{AB}/N = (N_{AB}/N_B)(N_B/N) = P(A \mid B)\, Pr(B) = P(B \mid A)\, Pr(A)$$

From which we obtain the Bayes theorem

$$P(A \mid B) = P(B \mid A)\, Pr(A)/Pr(B)$$

We can make several examples for the application of this formula. The first one is related to medical test to assess if based on the results of some clinical test you are sick or not. We take one suggested by Italian researcher on the AIDS. If the result of

G. Visconti, *Problems, Philosophy and Politics of Climate Science*,
Springer Climate, https://doi.org/10.1007/978-3-319-65669-4

the test is postive, you have a 100% probability to be infected. However, there exists a small probability that even if the test is positive you are not infected. Suppose that this correspond to a probability of 0.2%. This lower considerably the probability to be infected for a positive test. Suppose that the disease incidence is 1 Italian for each 600 (0.1666%) so if we make a test on 60 million of italian we will have 60 million/600 = 100.000 Italians. The remaining 59.9 million will give a number of false positive test 0.002 × 59.9 million = 119.800 false positive. The percentage of the people infected is then 100.000/(219.800) = 45.6%.

This very simple calculation can be put in Bayesian form, if we start defining the total probability of positive and negative test

$P(pos) = P(pos, AIDSyes) + P(pos, AIDSno)$

That can be written as conditional probability

$P(pos) = P(pos \mid AIDSyes) \times P(AIDSyes) + P(pos \mid AIDSno) \times P(AIDSno)$

This result can be used in the Bayes formula

$P(AIDSyes \mid pos) = P(pos \mid AIDSyes) \times P(AIDSyes)/P(pos)$

Substituting the values: $P(pos \mid AIDSyes) = 1$, $P(AIDSyes) = 1/600$, $P(pos \mid AIDSyno) = 0.002$ and $P(AIDSno) = 0.998$ we get

$P(AIDSyes \mid pos) = 45\%$

That is the same result we had before

The next example includes the improvement of the experimental data. We should define first the probability distribution of a variable F given a set of observation

$P(F \mid data) = P(data \mid F) \times P(F)/P(data)$

Again, we could apply this formula to a simple problem. Suppose that the temperature of two near localities (A and B) are similar. We know the series of the mean temperatures of the two localities and the distribution for A but not the distribution for B. We would like to determine the distribution parameter for B. The prior $P(F)$ is the distribution of the locality A. Combining the prior with the data average $P(data)$, we obtain the likelihood function $P(data \mid F)$. If we apply the Bayes formula we obtain the distribution at locality B, $P(F \mid data)$ that is our posterior. The process is particularly simple when the distribution is a gaussian, because the product of two gaussian is still a gaussian.

There is another way to use the Bayes statistics to find a parameter. The most stringent example refer to establishing if a coin is fair or not. We start by introducing a parameter (weight) that we call H. H = 0 indicates a money that gives always head; H = 0.5 will give a fair coin and H = 1 a coin that will give always tail. Applying Bayes we get

$P(H \mid data, I) = P(data \mid H, I) \times P(H, I)$

Where I indicates the hypothesis. The prior as a function of H can be taken as constant (the same value for all the H values), while the likelihood function in this case can be found based on a statistical which gives the number of ways in which r objects can be obtained once n is fixed. In this case when the number of flips (n) is fixed, the probability to have r heads is

$$P\,(dati \mid H) = {}^{n}C_r H_r (1 - H)^{n-r}$$

Where ${}^{n}C_r$ is a number which depends on both n and r. The distribution is the same as the binomial with success with probability H and failure with probability Q = 1 − H. In this case the experimental data on the statistics gives indications on the value of H. Bird (1998) gives and example in which three individuals A, B, and C have different "prejudices" on the coin. A think there is bias toward Head, B think the coin is fair and C that there is a bias toward tail. Analytically these hypothesis can be summarized as follows:

h_0 head probability less than 0.2

h_1 head probability between 0.2 and 0.4

h_2 head probability between 0.4 and 0.6

h_3 head probability between 0.6 and 0.8

h_4 head probability greater than 0.8

The prior probabilities attributed to the five hypotheses (h_0, h_1, h_2...) for the three individuals are reported in Table A.1. Actually these are not the probabilities $P\,(data \mid H)$ but rather the integral of this quantity that is

$$\int_{h_i}^{h_{i+1}} P(data \mid H)\,dH$$

where the interval $h_{i+1} - h_i = 0.2$.

At this point we can make the first test by trowing ten flips of the coin with the result of 10 heads. Then, we can calculate the distribution $P\,(data \mid H)$ shown on the right side of Fig. A.1. The gray area is proportional to the integral of the probability between H=0.2 and H=0.4 and is related to the probabilities h_0, h_1, h_2... of the previous table. We notice that at the increase of the experimental data the integral has a significative value only in the region between 0.2 and 0.4. We can calculate the Table A.2 after 100 flips. The value show that the coin is biased toward "tail" because the most probable value of H is between 0.2 and 0.4.

Table A.1 Prior probabilities for A, B, and C

	h_0	h_1	h_2	h_3	h_4
A	0.1	0.7	0.1	0.08	0.02
B	0.05	0.1	0.7	0.1	0.05
C	0.02	0.08	0.1	0.7	0.1

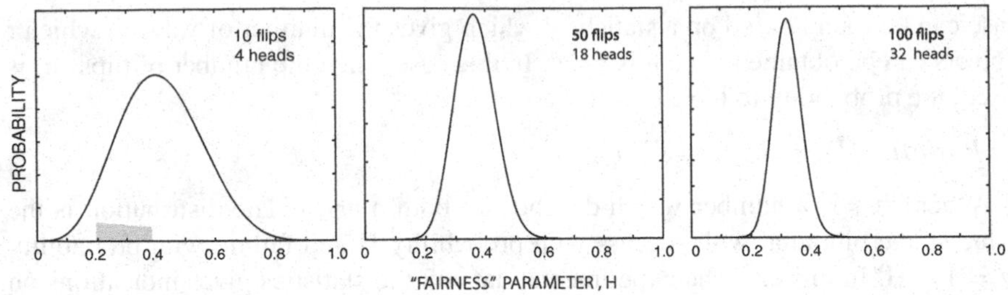

Fig. A.1 The Bayes probability for a coin tested for the fairness parameter H. H changes between 0 (always tails) and 1 (always heads). Notice that as the number of flips increase the maximum value of the function also increases

Table A.2 Probabilities after 100 flips

	h_0	h_1	h_2	h_3	h_4
A	0.0	0.999	0.01	0.0	0.0
B	0.0	0.948	0.052	0.0	0.0
C	0.0	0.990	0.001	0.0	0.0

A.2 The Test of the Hypothesis

Before going to the next example, we should introduce some additional concepts to the Bayes inference. We know the relationship between the prior $P(H_1)$ of some hypothesis H_1 to the posterior $P(H_1 \mid E)$ that utilizes the result of an experiment E. This can be done through the likelihood function $P(E \mid H_1)$, but also through the probability that the test be positive independently from the hypothesis H_1, $P(E)$ that is

$$P(H_1 \mid E) \times P(E) = P(E \mid H_1) \times P(H_1)$$

If we consider another hypothesis H_2 we could write a similar relation

$$P(H_2 \mid E) \times P(E) = P(E \mid H_2) \times P(H_2)$$

And if we make the ratio we get

$$P(H_1 \mid E)/P(H_2 \mid E) = P(E \mid H_1) \times P(H_1)/[P(E \mid H_2) \times P(H_2)]$$

Now the probability $P(E)$ must be expressed as a function $P(E \mid H)$, but also of the conditioned likelihood $P\left(E \mid \hat{H}\right)$. This can be done considering that the sum

$$P(H) + P(\hat{H}) = 1$$

Where $P(\hat{H})$ indicates that the test can be positive even if the starting hypothesis is wrong. So we write the two relations

$$P(H \mid E)\, P(E) = P(E \mid H)\, P(H);\; P\left(\hat{H} \mid E\right) P(E) = P\left(E \mid \hat{H}\right) P(\hat{H}),$$

that the test being positive even if the starting hypothesis is wrong. We have then

$$P(E) = P(E \mid H)\, P(H) + P\left(E \mid \hat{H}\right) P(\hat{H})$$

Substituting in the previous relation and consider that $P(\hat{H}) = (1 - P(H))$ we get

$$P(H \mid E) = P(E \mid H)\, P(H)/[P(E \mid H)\, P(H) + P\left(E \mid \hat{H}\right)(1 - P(H))]$$

We can use the above relation following an example made by Nate Silver in his book, of *Signal and the Noise* (Silver 2012). He considers the terrorist attack at the Twin Towers and assumes a very low probability for a terrorist attack before the first plane hits. After the first and second plane hits, he updates the probability and comes up with a number very near 1. He assume that the probability of a terrorist plane attack $P(TPA)$ is very low 1/20000 while the probability of an accidental plane crash $P(APC)$ is little bit higher 1/12500. We may also assume that $P(APC) = P\left(PC \mid T\hat{P}A\right)$, where $P(PC)$ is the probability of a plane crash and that APC and TPA are independent events. If we apply the just found formulation of the Bayes theorem

$$P(TPA \mid PC) = \frac{P(PC|TPA)P(TPA)}{[P(PC|TPA)P(TPA)+P\left(PC|T\hat{P}A\right)(1-P(TPA))]}$$

Substituting the numerical values we come up with

$$P(TPA \mid PC) = \frac{0.00005}{0.00005+0.00008(1-0.00005)} = 0.385$$

This constitutes the posterior probability after the first plane attack and if it is used as a new prior, we arrive to the probability of a terrorist attack

$$P(TPA \mid PC) = \frac{1\times0.385}{1\times0.385+0.00008(1-0.385)} = 0.9999$$

Some had objected to this calculation because for the second attack it was used the same probability of an accidental plane crash. The result does not change appreciably.

A.3 Fingerprinting

A simple way to start talking about fingerprinting is to introduce some simple statistical concept. Suppose for example to measure the maximum temperature of two near localities A and B, T_A and T_B. We can find the average of these two temperatures that we indicate as $\bar{T}_A$ and $\bar{T}_B$ and then we can calculate the deviation from the mean and the standard deviation σ_{T_A} and σ_{T_B} defined as for the deviations

$$\Delta T_A = T_A - \bar{T}_A$$

and for the standard deviations

$$\sigma_{T_A} = \sqrt{[(\Delta T_A)_1^2 + (\Delta T_A)_2^2 + ...(\Delta T_A)_N^2]/(N-1)}$$

where N is the total number of data.

These data can now be plotted as in Fig. A.2 where other two quantities are indicated the covariance and the correlation. The covariance between two sets of data is defined as

$$cov(T_A, T_B) = [(\Delta T_A \Delta T_B)_1 + (\Delta T_A \Delta T_B)_2 + \ldots (\Delta T_A \Delta T_B)_N]/(N-1)$$

while the correlation coefficient is defined as

$$corr(T_A, T_B) = cov(T_A, T_B)/\sigma_{T_A}\sigma_{T_B}$$

Before discussing the figure we can make a numerical example, where we assume the temperatures measured at site A is $T_A = 10, 12, 14, 18$ while $T_B = 9, 10, 16, 20$. The average for site A is 13.5 with deviations from the means -3.5, -1.5, 0.5 and 4.5. For site B the average is 13.75 and the deviations -4.75, -3.75, 2.25, 6.25. The standard deviations are for the two sites $\sigma_{T_A} = 3.41$ and $\sigma_{T_B} = 5.19$. It is easy to evaluate the sum of the square of the products and then the covariance that results 17.166. As a consequence the correlation coefficient is 0.97. This indicates that at the two sites the temperature increases in the same fashion and the covariance is positive. On the other hand, if we chose for one site a different behavior, for example, for site B a sequence like 19, 15, 9, 6 the covariance becomes negative (-19.66) and the correlation coefficient also -0.96.

Figure A.2 summarizes all these in a very simple way. At the top left, the two variables are independent and their covariance is near zero as well as the correlation coefficient. At the top right, the two variables show some degree of correlation (to and increase of A there is a correspondent increase in B), while the second line in the left the two are anticorrelated (one increases while the other decreases). Finally, at the lower right the correlation is almost perfect and the correlation coefficient is almost 1.

The last case is never observed in practice while the data most of the time are grouped in cloud that we exemplify with the gray ellipses. The noise is confined in the lower ellipse while in presence of a signal indicated by the arrow the signal added to the noise move the ellipse on the right and up. The method of the optimal detection indicated a direction near the real signal but that did not coincide with it.

Fig. A.2 Correlation, covariance, and ellipses of distribution. On *top* the possible correlation between data are shown while at the *bottom* the cartoon simplifies the optimal detection

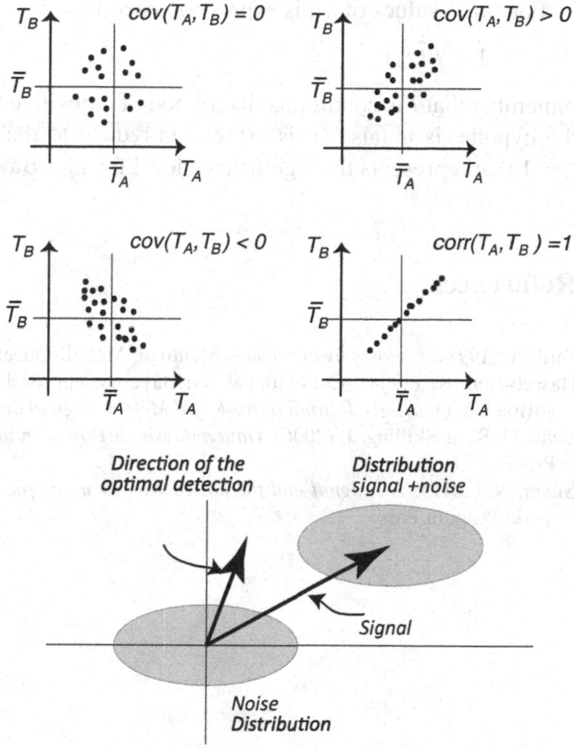

As for the comparison of the Bayes method with the fingerprinting, we can start from the Bayes theorem following Hasselmann (1998)

$$p\,(h \mid e) \times p(e) = p\,(e \mid h) \times p(h)$$

then write the net probability $p(e)$ as

$$p(e) = p\,(e \mid h)\,p(h) + p\left(e \mid \overline{h}\right)\,p(\overline{h}) = lc_0 + \widehat{l}(1 - c_0)$$

where we have used the definitions $\overline{l} = p\left(e \mid \overline{h}\right)$ and $c_0 = p(h)$.

And the Bayes theorem becomes

$$c = c_0 l/[c_0 l + (1 - c_0)\widehat{\overline{l}}]$$

and introducing the parameter

$$\beta = (1 - c_0)/c_0$$

$$c = (1 + \beta\widehat{\overline{l}}/l)^{-1}$$

For small values of $\widehat{l}$ this expression becomes

$$c \approx (1 - \beta\widehat{l}/l)$$

remember that $\widehat{l}$ is conditional likelihood of a positive test outcome for the case that the hypothesis is false. This expression reduce to the simpler one $c = 1 - \widehat{l}$ when $l \approx 1$ that represents the significance level for non Bayesian approach.

References

Bird, A. (1998). *Philosophy of science*. Montreal: McGill-Queen's University Press.

Hasselmann, K. (1998). Conventional and Bayesian approach to climate-change detection and attribution. *Quarterly Journal of the Royal Meteorological Society*, *124*, 2541–65.

Sivia, D. S., & Skilling, J. (2006). *Data anlaysis: A Bayesian tutorial*. Oxford: Oxford University Press.

Silver, N. (2012). *The signal and the noise: Why so many predictions fail–but some don't*. New York: Penguin Press.

Index

G. Visconti, *Problems, Philosophy and Politics of Climate Science*, Springer Climate, https://doi.org/10.1007/978-3-319-65669-4

Printed in the United States
By Bookmasters